D1217648

ege Library

S.S. Chern

A Great Geometer of the Twentieth Century

Expanded Edition

Edited by

S.-T. Yau

ISBN #: 1-57146-098-5

Typeset using LaTex
Printed on acid-free paper, in the United States of America.

Thus mathematics may be defined as the subject in which we never know what we are talking about nor whether what we are saying is true.

Bertrand Russell
(quoted by Chern)

S. S. Chern

Chern at the age of 5 in 1916 and his father.

Chern's family in 1930.

Chern in 1936.

Chern, Mrs Chow and W. L. Chow, Hamburg, 1936.

Buo Lung and May Chern, 1949.

Cherns at the University of Chicago in the 50s.

Cherns in the 50s.

Chern, S. Eilenberg and F. Hirzebruck, Mexico, 1956.

1968
Photograph taken by George Bergman of Berkeley.

1974
Photograph taken by George Bergman of Berkeley.

1977
Photograph taken by George Bergman of Berkeley.

1978

1979
Photograph taken by George Bergman of Berkeley.

1983
Photograph taken by George Bergman of Berkeley.

Deng Xiaoping and Chern, Beijing, 1984.

At the reception on the establishment of the Mathematical Sciences Research Institute at Berkeley, 1982.
(C. Moor, E. Thomas, I. Singer, Spanier, Heyman, Chern).

Wolf's Prize 83-84, presented by the President of Isreal.

C. Terng and Chern, 1985.

Chern received an honorary D. Sc. from Stony Brook, 1985.
Picture taken with C. N. Yang.

At the Institute Berkeley, 1985.

Chern & Mrs Chern, Helgason & Mrs Helgasn and Palais, 1987.

1988
Photograph taken by George Bergman of Berkeley.

At the Nankai Institute.

Chern and T. Y. Wu, Taipei, 1987.

At the Nankai Institute.

Paul Chu (son-in-law).

Lecturing at Nankai.

At the Nankai Institute.

Chern with his son Buo Lung and son-in-law Paul Chu, 1990.

At Nankai.

The Cherns in 1990.

Chern and S. T. Yau, 1992.

S. Y. Cheng, Chern and S. T. Yau, 1992.

Chern and S. T. Yau, 1992.

C.N. Yang, Mrs Chern and Chern, Taipei, 1992.

R. Baxter, I. Singer and D. Gross, 1992.

Chern, Mrs Chern, C. N. Yang and Mrs Yang
at the Conference in Honor of Yang's 70th birthday 1992.

Chern and I. M. Singer, 1992.

Cherns, Yangs, S. C. C. Ting and R. Baxter, 1992.

Lecturing at the conference in honor of C. N. Yang, 1992.

H. Hopf
J. Thompson

Table of Contents

Table of Contents, Continued

Table of Contents, Continued

Preface

In the summer of 1990, S.Y. Cheng and I organized a conference in Los Angeles in honor of our teacher Professor S.S. Chern on the occasion of his seventy ninth birthday. A large group of his friends and students gathered together for his birthday celebration. We reminisced about our experiences with our teacher. The lectures reflected the wisdom of this great mathematician and his warmth in interacting with young geometers.

Since the participants of the Symposium on differential geometry had already decided to dedicate the symposium to Professor Chern, we decided not to have any technical talks at the conference.

The purpose of this book is to record some of these talks. It also includes recollections of some of the distinguished scientists who were not able to come to conference. Some of the authors contributed technical essays. Since these papers reflect Professor Chern's mathematical interest, I decided to include them. I hope that through this book, the readers might get a glimpse of the life of a great geometer. I also take this opportunity to thank H.F. Lai and J. Simons for contributing funding to the conference for the purpose of inviting former students from China.

My Mathematical Education

Shiing-Shen Chern

1. Early Education in China

I entered Fulun Middle School in Tientsin in January 1923. It was a four-year high school and I was admitted to the second semester of the first year. The mathematical curriculum consisted of:

(1) First year, arithmetic, with textbook in Chinese;

(2) Second year, algebra, textbook by Hall and Knight;

(3) Third year, geometry, textbook by Wentworth and Smith;

(4) Fourth year, trigonometry and higher algebra, textbooks respectively by Wentworth-Smith and Hall-Knight.

My teachers were competent and devoted, and I did a large number of exercises. In the fourth year I was able to do many of the Cambridge-Tripos problems quoted in Hall-Knight's book.

I graduated from Fulun in 1926. In entering Nankai University the same year I skipped two years. As a result I never had a course on analytic geometry. It was worse than that: I had to take an entrance examination to Nankai whose mathematical part consisted principally of analytic geometry. For three weeks before the examination I studied by myself the book "Mathematical Analysis" by Young and Morgan.

If I remember correctly, I came out with the second best paper of the examination. The notion of "foci of conics" puzzled me for a long time until I learned projective geometry several years later.

After entering Nankai University I soon found that I was clumsy with experimental work and mathematics became my only choice. I was fortunate to find as teacher Professor Li-Fu Chiang, who received his Ph.D. from Harvard in 1918 with Julian Coolidge, with a thesis on the line-sphere contact transformation in a noneuclidean space. As a result I did a lot of study on geometry during my senior year. Among the books I studied were Coolidge's Non-euclidean Geometry, Geometry of the Circle and Sphere, Salmon's Conic Sections and Analytic Geometry of Three Dimensions, and Castelnuovo's Analytic and Projective Geometry, etc. I was particularly fascinated by Otto Staude's two-volume book on "Fadenkonstruktionen." The geometry of quadrics is a beautiful chapter of mathematics. I was glad to see it taken up by J. Moser in 1979 in his work on integrable Hamiltonian systems and spectral theory; cf.[3] of Bibliography. Even now a study of "Salmon" could be rewarding, and is in my opinion enjoyable.

After graduation from Nankai in 1930 I went to work with Professor Dan Sun of Tsing Hua University in Peiping. He was at that time the only mathematician in China publishing research papers on mathematics. His field was projective differential geometry, being a Ph.D. student of E.P. Lane at the University of Chicago. The subject was founded by E.J. Wilczynski in 1901 and was a natural outgrowth of projective geometry which had reigned over geometry for almost a century. I became familiar with the literature and wrote a few papers. Among them was my master's thesis on projective line geometry. Following Plücker and Klein line geometry had been a favorite topic of geometers. In fact, Klein's dissertation was on quadratic line complexes, i.e., line loci defined by quadratic equations in the Plücker coordinates. They have beautiful properties; a modern treatment can be found in the book of

Griffiths-Harris [1]. Line geometry is very much in the background of twistor theory.

My thesis studies line congruences, i.e., two-dimensional submanifolds of lines, and their osculation by quadratic line complexes.

Toward the end of my graduate years, i.e., around 1934, I began to realize the importance of global differential geometry, called differential geometry in the large at that time. It was generally considered to be a difficult subject, both in the mathematical breadth required and in the depth of the problems. My main inspiration came from Wilhelm Blaschke's books on differential geometry.

It was clear that algebraic topology was at the foundation of the whole area. But algebraic topology itself was then in a stage of development. Veblen's "analysis situs", published in 1922, introduced the "homology characters", i.e., Betti numbers and torsion coefficients, in terms of the incidence matrices. Lefschetz's Topology appeared in 1930, but it did not help the beginners to enter the subject. I had a course (1933-34) from Emanuel Sperner, then visiting at Peking University, where he gave, among other things, a careful and detailed account of Erhard Schmidt's proof of the Jordan curve theorem. I also had a course with Tsai-Han Kiang, a former student of Marston Morse and a former assistant to Lefschetz, on analysis situs, following Lefschetz's book. But I had the feeling that I was only at the door of the great edifice of algebraic topology. The situation changed greatly only with the appearance of the Seifert-Threlfall book in 1934 and the Alexandroff-Hopf book in 1935.

In the spring of 1932 Blaschke visited Peiping and gave a series of lectures on "topological questions in differential geometry". It was really local differential geometry where he took, instead of a Lie group as in the case of classical differential geometries, the pseudo-group of all diffeomorphisms and studied the local invariants. I was able to follow his lectures and to read many papers under the same general title published in the Hamburger Abhandlungen and other journals. The subject is now known as web geometry. With this contact and my previous knowledge of Blaschke's books on differential geometry, I decided to go to Hamburg as a student when a fellowship was made available to me in 1934.

2. European Student Life

I was in Hamburg in 1934-36, receiving my D.Sc. in 1936, and spent a postdoctoral year in Paris with Elie Cartan. The choice of Hamburg turned out to be a fortunate one. Hamburg had a strong Department, with professors Blaschke, Artin, and Hecke, and junior members including E. Kähler, H. Petersson, H. Zassenhaus.

Blaschke's mathematical interest was shifting from web geometry to integral geometry. When I first saw him in September 1934, he gave me a bunch of reprints on web geometry. I became interested in the notion of the rank of a web and webs of maximum rank. Recall that a d-web in R^n of codimension 1 consists of d foliations by hypersurfaces in general position. If $x_1, \cdots, x_n$ are the coordinates in R^n and the foliations given by the equations

$$u_i(x_1, \cdots, x_n) = \text{const}, 1 \leq i \leq d,$$

an equation of the form

$$\sum_{1 \leq i \leq d} f_i(u_i) = 0$$

is called an abelian equation. The maximum number of linearly independent abelian equations is called the rank of the web. If the d-web is defined by the hyperplanes of an algebraic curve of class d in R^n, it has abelian equations given by Abel's theorem applied to the abelian differentials. Hence its rank is at least the genus of the curve. In a short note I determined the maximum rank $\pi(d, n), n \leqq d-1$, of all codimension one d-webs in R^n. According to a theorem of Castelnuovo this integer is equal to the maximum genus of an algebraic curve of degree d in the projective space P^n of dimension n, which does not belong to a hyperplane, P^{n-1}. The remarkable fact is that not all maximum rank webs are given by maximum genus algebraic curves in the manner described above: there are exotic maximum rank webs, whose leaves are not all hyperplanes. The abelian equations are essentially functional equations, for in the classical cases they become addition theorems of well-known transcendental functions. In the plane ($n = 2$) a 5-web of curves has maximum rank 6 and there exists an exotic web (Bol's web) whose abelian equations involve the dilogarithm. Griffiths and I studied in 1978 the question of codimension one d-webs in R^n of maximum rank $\pi(d, n)$, but we did not reach the goal. I think the determination of such webs which are exotic is a problem of great interest and importance.

During 1934-35 my major effort was spent on Kähler's seminar. It was based on his famous booklet "Einführung in die Theorie der Systeme von Differentialgleichungen", which had just been published. The main result was later known as the Cartan-Kähler Theorem. At the first meeting all the people were present, including Blaschke, Artin, and Hecke, and everybody was given a copy of the book. The attendance decreased rapidly and I was one of very few who stayed till the end. I made an application of the theory to 3-webs of r-dimensional submanifolds in R^{2r}. Both Blaschke and Kähler thought this and my earlier result on maximum rank were enough for a thesis. So I had my thesis ready by the end of 1935.

Blaschke and his school were mainly concerned with integral geometry, on which he gave a course. The most beautiful results were found by L.A. Santalò. One result consists of expressing the isoperimetric defect of a plane convex curve as an infinite sum of positive terms each of which has a geometric meaning. Santalò went on to become a world leader on integral geometry. He was from Spain and immigrated to Argentina.

Another of my fellow students was the algebraic geometer Wei-Liang Chow. He came from Chicago to Göttingen in order to wrok with Hermann Weyl. Political developments in Göttingen and Germany made this impossible, and he went to work with van der Waerden in Leipzig. But for some reason he lived in Hamburg and occasionally took part in seminars. He was developing his "zugeordnete Formen" later known as Chow coordinates. Chow is an original mathematician. He made major contributions to algebraic geometry, including his theorem on compact subvarieties and his intersection theory. He came from a high mandarin family in China, which recognized early the need of westernization. As a result the family produced many distinguished people. Chow was a night worker. When he visited me, I lost some sleep but learned some mathematics.

I attended Artin's lectures whenever possible. During the two years they included: complex function theory, algebraic topology, relativity, and diophantine approximations. I also had a course with Hecke on algebraic number theory, following mostly his book. The scientific life in Hamburg was an ideal one, but political events did not allow it to last.

I had a postdoctoral year in 1936-37 and sought the opinion of Blaschke. He advised me either to stay on in Hamburg and work with Artin on number theory or to go to Paris to work with Elie Cartan. They were attractive alternatives, but Paris and Cartan won.

The timing was perfect. For in that year Cartan gave a course on exterior differential systems; the lecture notes later came out as a book. The "young" French mathematicians, who later became Bourbaki, began to be active. They organized a "Séminaire Julia", which met biweekly and was devoted to a topic to be chosen every year. the topic for 1936-37 was "Les travaux de M. Elie Cartan".

Cartan was a wonderful teacher. He suggested "little" problems, some of which became the subjects of my papers. Probably because of my responses to his questions he allowed me to visit him at his home, about once every two weeks. After the visit I usually received a letter from him the next day, which would say: "After your departure I thought more about your questions. ... It would be interesting ... ". It was an interesting and unforgettable year.

I also attended Montel's lectures on several complex variables and Hadamard's seminar at the Collége de France. At the end of a seminar Hadamard would give a summary, which was frequently more lucid and informative than the talk itself.

On July 10, 1937 I left Paris to return to China with a heavy heart, after learning the news that the Sino-Japanese war had broken out.

3. Mathematical Isolation

When I left Eruope for China in the summer of 1937, I was going to take up my position as professor of mathematics at Tsing Hua University in Peiping. Because of the Sino-Japanese war the goal was reached only ten years later. The University moved to Changsha and then to Kunming in 1938, until the end of the war in the summer of 1945.

Kunming is a beautiful city. With deprivations and uncertainties to be expected in a country at war, life was otherwise pleasant. Tsing Hua University joined with Peking University and Nankai University to form the Southwest Associated University, and Kunming immediately became the intellectual center of wartime China. My mathematics colleagues included Loo-keng Hua and Pao-lu Hsu. I gave classes and seminars on algebraic topology, Lie groups, sphere geometry, exterior differential systems, etc. and attracted a reasonable number of students. The great disadvantage was that the place was cut off from the outside: there was a period when even the "Burma Road" was closed and the only communication with the outside world was by air. I had a small personal library. At the beginning it was even fun to do some reading and thinking that I had wished and did not find the time. But frustrations came quickly and had to be overcome. I wrote to Elie Cartan telling him the situation and he sent me a large number of his reprints, including some old ones. I spent a great deal of time pondering over them and thinking about their implications and applications. This was definitely a gain. In the thirties people such as Weyl, Blaschke, and Kähler began to realize the importance of Cartan's work, but very few read his old papers (except those on Lie algebra). I was lucky to be forced to carry it out.

The Chinese ambassador in Washington, Dr. Hu Shih, sent by air mail a copy of the book by Hurewicz-Wallman on "Dimension theory". People now used to xerox may find it difficult to imagine that I copied the whole book by hand, with the exception of the last chapter, where they did "exact sequences" without the sequence and I found it difficult to follow. In fact, at that time it was common to take notes from the reading of a paper. In contrast to the xerox flood, it is not clear whether we have made progress.

I began to have students, among whom were Hsien-Chung Wang and Chih-ta Yen. Wang later made extensive contributions to topology, although he was best known for the Wang sequence. Yen was the first one to give the correct values of the Betti numbers of all the exceptional Lie groups.

Looking back I do not think I had a good idea of mathematics as a whole. I knew some of my deficiencies and was anxious to fill them. My mathematical strength lies in my ability in computation. Even now I do not mind doing lengthy computations, while years ago I could do them with relatively few errors. This is a training which is now relatively unpopular and has not been encouraged. It is still a great advantage in dealing with many problems.

I was fascinated by the Gauss-Bonnet formula and knew that the most conceptual proof is through the structural equation expressing the exterior derivative of the connection form. So when I went to Princeton in 1943, the ground was laid for a most satisfying piece of my mathematical work.

4. Princeton Sunshine

I arrived in Princeton in August 1943. The change of atmosphere was unforgettable. The Institute for Advanced Study was quiet during these days, as most people left for war work. Hermann Weyl was interested in my work. Before my visit he refereed a paper of mine on isotropic surfaces for the Annals of Mathematics and wrote a long favorable report, a fact he revealed to me personally. The report contained suggestions for improvements and showed that he went through the paper in detail. We had frequent conversations. Among his insights was the prediction that algebraic geometry was going to have a great future.

André Weil was at nearby Lehigh University. We soon met and had a lot to talk about. Weil had just published his paper with Allendoerfer on the Gauss-Bonnet formula, and it immediately entered into our discussions. From my understanding of the two-dimensional case I knew that the right proof had to be based on an idea that we now call transgression. There were two difficulties: 1) I did not know clearly the Poincaré-Hopf theorem on the singularities of a vector field; 2) the transgression has to be carried out in the unit tangent bundle and not in the principal bundle, which involves a non-trivial technical difficulty. These were overcome within a short time and the story had a happy ending. I still consider this my best piece of work.

After this it is natural to extend the result to the Stiefel-Whitney classes. That was the time when even in Princeton a talk on fiber bundles had to begin with a definition; there were no vector bundles, only sphere bundles. I noticed that the complex characteristic classes are simpler and admit a local curvature representation. The work was not difficult, but was not in the topology fashion at that time.

Although I was a member of the Institute, I spent a lot of time in Fine Hall of the University. Chevalley was writing his book on Lie groups. Lefschetz was opinionated and did not like the routine work on differential geometry then prevailing. When he asked me to referee a paper for the Annals and I recommended rejection, he made me an associate editor.

The surroundings and pace were most agreeable to me. I reached greater maturity in my mathematical outlook and I enjoyed the stay greatly. In recent years scientific competition has reached a proportion making the life of a scientist unpleasant, although the situation is much better with mathematics. I do not think there is a need for quick progress and I am not impressed by the discovery of electronic mail.

I left Princeton at the end of 1945, returning to China. Immediately on arrival I was given the task of organizing an Institute of Mathematics in in Academia Sinica, the Chinese National Academy. Although the second World War ended, China was torn by a civil war. I extended an invitation to Hermann Weyl to visit China. He responded favorably, but conditions in China made such a visit impossible.

At the end of 1948 the Nanking government was collapsing. I was grateful to the Institute for Advanced Study for taking the initiative to bring me out of China. I was at the Institute in the winter term of 1949. During the term I was the main speaker in Veblen's seminar on differential geometry. The notes were written up two years later and had a wide circulation; they are now published in volume IV of my "Selected Papers". The main result is the Weil homomorphism.

It is a generalization of the Chern classes from the unitary group to any Lie group. I knew the result while writing my paper on complex characteristic classes in 1944. I could not prove it, not being on top of Lie groups. Weil furnished a crucial idea, by considering a family of connections. I called the result the Weil homomorphism. My friends thought I should have a share of the credit, to which I naturally do not object.

5. Mathematics at the Midway

After the second world war Marshall Stone was called to reorganize the department of mathematics at the University of Chicago. He became chairman. As a proof of his insight in mathematics and the mathematical world his first two offers went respectively to Hassler Whitney and André Weil. Whitney declined, but after some negotiations Weil accepted.

When I was in China, Stone corresponded with me about a visiting appointment in Chicago. After I came in 1949, the Department decided to make me a permanent offer. I think the University of Chicago is the only American university whose main objective is the "advancement of knowledge" and not education. I had many friends in the Department and I joined it in summer of 1949. It turned out to be a very pleasant and profitable association.

In 1949-50 I gave a course entitled "differential geometry in the large" and I had a galaxy of brilliant students. I myself was finding my way; many of my errors and blunders were duly corrected by my students. It was an animated and interesting gathering. My memory goes to Arnold Shapiro, who led many of the discussions. Looking back, my knowledge of differential geometry was rudimentary. It is perhaps the strength of the subject that some issues are even now unsettled. For instance, what is a surface? Is it imbedded, immersed, or defined by equations with possible singularities? On the other hand, many of the topics touched in the class received extensive later developments.

I had a close association with Weil. He was always ready and available. Among the mathematicians I discussed mathematics with, and there were many, Weil was one of the few who grasped my ideas quickly and gave helpful comments. We took long walks along Lake Michigan when it was still safe.

I was also interested in algebraic topology, and occasionally taught a course. With Ed Spanier we did some joint work on sphere bundles. One of our results was to formulate Gysin's work as an exact sequence. René Thom did it in a cleaner way and the result is commonly known as the Thom isomorphism.

I found both Chicago and Hamburg very enjoyable. I think they are of the right size. Unfortunately developments in mathematics have forced everything to grow bigger.

6. Settlement on the West Coast

In 1960 I moved to Berkeley. The place was not unfamiliar to me. My teacher in China, Professor L.F. Chiang, received his B.S. from Berkeley. In 1946 and 1949 I stopped at Berkeley and spent some time with the Department. The Department was first-rate and was built up by G.C. Evans. On several occasions he asked about my interest in joining. His brother was the owner of the famous western bookstore in Tientsin, where I got some of my textbooks but was generally frightened by the prices.

Ironically the Berkeley offer became serious when Evans was retiring. It was true, as it was sometimes speculated, that I was attracted by the milder climate as I was getting older. But other factors, such as the expanding department and jet travel, making California less isolated, also favored the move.

Berkeley was improving her standing in the mathematical world and attracting excellent students. There were 31 students who got their Ph.D.'s with me, but my influence extended also to others. I began to write joint papers, with myself as the junior author, as in the case of Bott, Griffiths, Moser, Simons, etc. In such cases I had the feeling of a light responsibility. Life became more and more pleasant.

The colleagues with close scientific contact with me included Hans Lewy and Chuck Morrey, original and powerful analysts. Lewy and I spent some time on the problem of local isometric imbedding of a three-dimensional Riemannian metric in R^6. We were led to the cubic asymptotic cone and knew that it is hyperbolic, but stopped there.

The rôle of differentiation in mathematics is a mystery. One is inclined to think that the two pillars of mathematics are algebra and topology. But life is not that simple; Newton and Leibniz played a trick. This period saw the admission of differential geometry to the main stream of mathematics.

7. Something to Play With in My Eighties

My career is approaching an end and my only question is what to do. The answer is simple: I will continue to play with mathematics. I have never been alert in physical activities and it is now out of the question. Music I have found to be a waste of time. My occasional involvement is purely social. Fortunately global differential geometry still has many fundamental problems, although I most likely will be only a spectator in its development.

I think the restriction to smooth manifolds is dictated only by technical reasons and is unsatisfactory. Not only do non-smooth manifolds exist naturally. But also even if we start with a smooth manifold, geometrical constructions, such as the evolute, lead to non-smooth ones. Whitney introduced the notion of a stratified manifold, which allows singularities and the application of the infinitesimal calculus. Recent light was cast by the work of Robert McPherson. The Cheeger-Goresky-McPherson intersection homology and the McPherson Chern classes have given substance to the notion; cf [2].

It is also not clear to me whether the Riemannian structure is as basic as indicated by recent developments. After all, in his historical paper Riemann allowed the metric to be the fourth root of a quartic from, the general case being now called Finslerian. In a recent note [4] I showed that Finsler geometry can be developed simply, if the proper viewpoint is taken. Further developments are inevitable.

Due to my background I like algebraic manipulation, as Griffiths once observed. Local differential geometry calls for such work. But good local theorems are difficult to come by. The problem on maximum-rank webs discussed above is clearly an important problem, and will receive my attention.

My mathematical education goes on.

References

1. P. Griffiths and J. Harris, *Principles of Algebraic Geometry*, John Wiley, 1978.

2. Robert McPherson, *Global questions in the topology of singular spaces, Proc. ICM Warszawa*, vol. 1, 1983, pp. 213-235.

3. J. Moser, *Geometry of quadrics and spectral theory, Chern Symposium*, Springer-Verlag, 1979, pp. 147-188.

4. S. Chern, *On Finsler geometry*, Comptes Rendus, Académie des Sciences, Paris t. 314, 1992, p.p. 757-761.

New York, le 20 juillet 1937

Mon cher maître,

D'abord je veux vous adresser mes remerciements cordials et ma reconnaisance profonde pour vos intérêts et pour toutes les aides que vous m'avez données dans mes études. Pendant mon court séjour à Paris vous m'avez indiqué un champ important de recherches qui est encore peu exploré et que j'espère d'étudier en Chine.

Je suis arrivé à New York le 15 juillet. J'ai visité l'Université à Princeton, où j'ai eu un entretien fort interessant avec M. T. Y. Thomas. M. Thomas m'a parlé de ma Note et il m'a dit qu'il ~~lui~~ vous a prié de la lui envoyer. Je continuerai bientôt mon voyage pour la Chine. Je vous écrirai plus tard.

Veuillez agréer, mon cher maître mes plus profondes considérations.

S. S. Chern

HENRI CARTAN
95, BOULEVARD JOURDAN
F - 75014 PARIS
TÉL. (1) 45.40.51.78

Paris, 27 juin 1990

Cher Professeur Chern,

Je regrette de ne pas être présent le 14 juillet 1990 lorsque vous recevrez l'hommage de vos amis, collègues et disciples. C'est de tout coeur que je m'associe par la pensée à cet hommage. Vous êtes pour moi le continuateur de l'oeuvre d'Elie Cartan et le créateur qui a renouvelé la Géométrie différentielle globale dans ses rapports avec la Topologie.

Bien fidèlement à vous.

H. Cartan

P.S.- J'espère que les copies ci-jointes de deux lettres que vous avez adressées à mon père en 1936 et 1937 vous intéresseront.

The Life and Mathematics of Shiing-Shen Chern

Dedicated to S.S. Chern for the celebration of his 79th Birthday

Richard S. Palais and Chuu-Lian Terng

Introduction

Many mathematicians consider Shiing-Shen Chern to be the outstanding contributor to research in differential geometry in the second half of the twentieth century. Just as geometry in the first half-century bears the indelible stamp of Élie Cartan, so the seal of Chern appears large on the canvas of geometry that has been painted in the past fifty years. And beyond the great respect and admiration that his scientific accomplishments have brought him, there is also a remarkable affection and esteem for Chern on the part of countless colleagues, students, and personal friends. This reflects another aspect of his career – the friendship, warmth, and consideration Chern has always shown to others throughout a life devoted as much to helping younger mathematicians develop their full potential as to his own research.

Our recounting of Chern's life is in two sections: the first, more biographical in nature, concentrates on details of his personal and family history; the second gives a brief report on his research and its influence on the development of twentieth-century mathematics.

Our main sources for the preparation of this article were the four volumes of Chern's selected papers [CSP] published by Springer-Verlag, a collection of Chern's Chinese articles by Science Press [SWC], and many conversations with Chern himself. Letters within square brackets refer to the references at the end of this article, whereas numbers within square brackets refer to items from the *Bibliography of the Publications of S.S. Chern*, found in the second volume of [CSP].

Early life

Chern was born on October 28, 1911 in Jia Xin. His father, Bao Zheng Chern, passed the city level Civil Service examinations at the end of the Qing Dynasty, and later graduated from Zhe Jiang Law School and practiced law. He and Chern's mother, Mei Han, had one other son and two daughters.

Because his grandmother liked to have him at home, Shiing-Shen was not sent to elementary school, but instead learned Chinese at home from his aunt. His father was often away working for the government, but once when his father was at home he taught Shiing-Shen about numbers, and the four arithmetic operations. After his father left, Shiing-Shen went on to teach himself arithmetic by working out many exercises in the three volumes of Bi Shuan Mathematics. Because of this he easily passed the examination and entered Xiu Zhou School, fifth grade, in 1920.

His father worked for the court in Tianjin and decided to move the family there in 1922. Chern entered Fu Luen middle school that year and continued to find mathematics easy and interesting. He worked a large number of exercises in *Higher Algebra* by Hall and Knight, and in *Geometry and Trigonometry* by Wentworth and Smith. He also enjoyed reading and writing.

1926-30, Nankai University

Chern passed the college entrance examinations in 1926, at age fifteen, and entered Nankai University to study Mathematics. In the late 1920's there were few mathematicians with a PhD degree in all of China, but Chern's teacher, Lifu Jiang, had received a doctoral degree from Harvard with Julian Coolidge. Jiang had a strong influence on Chern's course of study; he was very serious about his teaching, giving many exercises and personally correcting all of them. Nankai provided Chern with an excellent education during four happy years.

1930-34, Qing Hua graduate school

In the early 1930's, many mathematicians with PhD degrees recently earned abroad were returning to China and starting to train students. It appeared to Chern that this new generation of teachers did not encourage students to become original and strike out on their own, but instead set them to work on problems that were fairly routine generalizations of their own thesis research. Chern realized that to attain his goal of high quality advanced training in mathematics he would have to study abroad. Since this family could not cover the expense this would involve, he knew that he would require the support of a government fellowship. He learned that a student graduating from Qing Hua graduate school with sufficiently distinguished records could be sent abroad with support for further study, so, after graduating from Nankai in 1930, he took and passed the entrance examination for Qing Hua graduate school. At that time the four professors of mathematics at Qing Hua were Qinglai Xiong, Guangyuan Sun, Wuzhi Yang (C.N. Yang's father), and Zhifan Zheng (Chern's father-in-law to be), and Chern studied projective differential geometry with Professor Sun

While at Nankai Chern had taken courses from Jiang on the theory of curves and surfaces, using a textbook written by W. Blaschke. Chern had found this deep and fascinating, so when Blaschke visited Beijing in 1932, Chern attended all of his series of six lectures on web geometry. In 1934, when Chern graduated from Qing Hua, he was awarded a two-year fellowship for study in the United States but, because of his high regard for Blaschke, he requested permission from Qing Hua to use the fellowship at the University of Hamburg instead. The acting chairman, Professor Wuzhi Yang, helped both to arrange the fellowship for Chern and for his permission to use it in Germany. This was the year that the Nazis were starting to expel Jewish professors from the German universities, but Hamburg University had opened only several years before and, perhaps because it was so new, it remained relatively calm and a good place for a young mathematician to study.

1934-36, Hamburg University

Chern arrived at Hamburg University in September of 1934, and started working under Blaschke's direction on applications of Cartan's methods in differential geometry. He received the Doctor of Science degree in February 1936. Because Blaschke travelled frequently, Chern worked much of the time with Blaschke's assistant, Kähler. Perhaps the major influence on him while at Hamburg was Kähler's seminar on what is now known as Cartan-Kähler Theory. This was then a new theory and everyone at the Institute attended the first meeting. By the end of the seminar only Chern was left, but he felt that he had benefited greatly from it.

When his two year fellowship ended in the summer of 1936, Chern was offered appointments at both Qing Hua and Beijing University. But he was also offered another year of support from The Chinese Culture Foundation and, with the recommendation of Blaschke, he went to Paris in 1936-37 to work under the renowned geometer Élie

Cartan.

1936-37, Paris

When Chern arrived in Paris in September of 1936, Cartan had so many students eager to work with him that they lined up to see him during his office hours. Fortunately, after two months Cartan invited Chern to see him at home for an hour once every other week during the remaining ten months he was in Paris. Chern spent all his efforts preparing for these biweekly meetings, working very hard and very happily. He learned moving frames, the method of equivalence, more of Cartan-Kähler theory, and most importantly according to Chern himself, he learned the mathematical language and the way of thinking of Cartan. The three papers he wrote during this period represented the fruits of only a small part of the research that came out of this association with Cartan.

1937-43, Kunming and The Southwest University Consortium

Chern received an appointment as Professor of Mathematics at Qing Hua in 1937. But before he could return to China, invading Japanese forces had touched off the long and tragic Sino-Japanese war. Qing Hua joined with Peking University and Nankai University to form a three-university consortium, first at Changsha, and then, beginning in January 1938, at Kunming, where it was called the Southwest Associated University. Chern taught at both places. It had an excellent faculty, and in particular Luogeng Hua was also Professor of Mathematics there. Chern had many excellent students in Kunming, some of whom later made substantial contributions to mathematics and physics. Among these were the mathematician H.C. Wang and the Nobel prize-winning physicist C.N. Yang. Because of the war, there was little communication with the outside world and the material life was meager. But Chern was fortunate enough to have Cartan's recent papers to study, and he immersed himself in these and in his own research. The work begun during this difficult time would later become a major source of inspiration in modern mathematics.

Chern's family

In 1937 Chern and Ms. Shih-Ning Cheng became engaged in Changsha, having been introduced by Wuzhi Yang. She had recently graduated from Dong Wu University, where she had studied biology. They were married in July of 1939, and Mrs. Chern went to Shanghai in 1940 to give birth to their first child, a son Buo Lung. The war separated the family for six years and they were not reunited until 1946. They have a second child, a daughter, Pu (married to Chingwu Chu, one of the main contributors in the development of high temperature superconductors).

The Cherns have had a beautiful and full marriage and family life. Mrs. Chern has always been at his side and Chern greatly appreciated her efforts to maintain a serene environment for his research. He expressed this in a poem he wrote on her sixtieth birthday:

Thirty-six years together
Through times of happiness
And times of worry too.
Time's passage has no mercy.

We fly the Skies and cross the Oceans
To fulfill my destiny;
Raising the children fell
Entirely on your shoulders.

How fortunate I am
To have my works to look back upon,

I feel regrets you still have chores.

Growing old together in El Cerrito is a blessing.
Time passes by,
And we hardly notice.

In 1978 Chern wrote in the article "A summary of my scientific life and works":

> "I would not conclude this account without mentioning my wife's role in my life and work. Through war and peace and through bad and good times we have shared a life for forty years, which is both simple and rich. If there is credit for my mathematical works, it will be hers as well as mine."

1943-45, Institute for Advanced Study at Princeton

By now Chern was recognized as one of the outstanding mathematicians of China, and his work was drawing international attention. But he felt unsatisfied with his achievements, and when O. Veblen obtained a membership for him at the Institute for Advanced Study in 1943, he decided to go despite the great difficulties of wartime travel. In fact, it required seven days for Chern to reach the United States by military aircraft!

This was one of the most momentous decisions of Chern's life, for in those next two years in Princeton he was to complete some of his most original and influential work. In particular, he found an intrinsic proof of The Generalized Gauss-Bonnet Theorem [25], and this in turn lead him to discover the famous Chern characteristic classes [33]. In 1945 Chern gave an invited hour address to the American Mathematical Society, summarizing some of these striking new advances. The written version of this talk [32] was an unusually influential paper; and as Heinz Hopf remarked in reviewing it for *Mathematical Reviews* it signaled the arrival of a new age in global differential geometry" (Dieser Vortrag...zeigt, dass wir uns einer neuen Epoche in der "Differentialgeometrie im Grossen" befinden).

1946-48, Academia Sinica

Chern returned to China in the spring of 1946. The Chinese government had just decided to set up an Institute of Mathematics as part of Academia Sinica. Lifu Jiang was designated chairman of the organizing committee, and he in turn appointed Chern as one of the committee members. Jiang himself soon went abroad, and the actual work of organizing the Institute fell to Chern. At the Institute, temporarily located in Shanghai, Chern emphasized the training of young people. He selected the best recent undergraduates from universities all over China and lectured to them twelve hours a week on recent advances in topology. Many of today's outstanding Chinese mathematicians came from this group, including Wenjun Wu, Shantao Liao, Guo Tsai Chen, and C.T. Yang. In 1948 the Institute moved to Nanjing, and Academia Sinica elected eighty-one charter members, Chern being the youngest of these.

Chern was so involved in his research and with the training of students that he paid scant attention to the civil war that was engulfing China. One day however, he received a telegram from J. Robert Oppenheimer, then Direction of the Institute for Advanced Study, saying "If there is anything we can do to facilitate your coming to this country please let us know." Chern went to read the English language newspapers and, realizing that Nanjing would soon become embroiled in the turnmoil that was rapidly overtaking the country, he decided to move the whole family to America. Shortly before leaving China he was also offered a position at the Tata Institute in Bombay. The Cherns left from Shanghai on December 31, 1948, and spent the Spring Semester at the Institute in Princeton.

1949-60, Chicago University

Chern quickly realized that he would not soon be able to return to China, and so would have to find a permanent position abroad. At this time, Professor Marshall Stone of the University of Chicago Mathematics Department had embarked on an aggressive program of bringing to Chicago stellar research figures from all over the world, and in a few years time he had made the Chicago department one of the premier centers for mathematical research and graduate education worldwide. Among this group of outstanding scholars was Chern's old friend, André Weil, and in the summer of 1949 Chern too accepted a professorship at the University of Chicago. During his eleven years there Chern had ten doctoral students. He left in 1960 for the University of California at Berkeley, where he remained until his retirement in 1979.

Chern and C.N. Yang

Chern's paper on characteristic classes was published in 1946 and he gave a one semester course on the theory of connections in 1949. Yang and Mills published their paper introducing the Yang-Mills theory into physics in 1954. Chern and Yang were together in Chicago in 1949 and again in Princeton in 1954. They are good friends and often met and discussed their respective research. Remarkably, neither realized until many years later that they had been studying different aspects of the same thing!

1960-79, UC Berkeley

Chern has commented that two factors convinced him to make the move to Berkeley. One was that the Berkeley department was growing vigorously, giving him the opportunity to build a strong group in geometry. The other was...the warm weather.

During his years at Berkeley, Chern directed the thesis research of thirty-one students. He was also teacher and mentor to many of the young postdoctoral mathematicians who came to Berkeley for their first jobs. (This group includes one of the coauthors of this article; the other was similarly privileged at Chicago.)

During this period the Berkeley Department became a world-famous center for research in geometry and topology. Almost all geometers in the United States, and in much of the rest of the world too, have met Chern and been strongly influenced by him. He has always been friendly, encouraging, and easy to talk with on a personal level, and since the 1950's his research papers, lecture notes, and monographs have been the standard source for students desiring to learn differential geometry. When he "retired" from Berkeley in 1979, there was a week long "Chern Symposium" in his honor, attended by over three hundred geometers. In reality, this was a retirement in name only; during the five years that followed, not only did Chern find time to continue occasional teaching as Professor Emeritus, but he also went "up the hill" to serve as the founding director of the Berkeley Mathematical Sciences Research Institute (MSRI).

1981-present. The Three Institutes

In 1981 Chern, together with Calvin Moore, Isadore Singer, and several other San Francisco Bay area mathematicians wrote a proposal to the National Science Foundation for a mathematical research institute at Berkeley. Of the many such proposals submitted, this was one of only two that were eventually funded by the NSF. Chern became the first director of the resulting Mathematical Sciences Research Institute (MSRI), serving in this capacity until 1984. MSRI quickly became a highly successful institute and many credit Chern's influence as a major factor.

In fact, Chern has been instrumental in establishing three important institutes of mathematical research: The Mathematical Institute of Academia Sinica (1946), The Mathematical Sciences Research Institute in Berkeley, California (1981), and The Nankai Institute for Mathematics in Tianjin, China (1985). It was remarkable that Chern did this despite a reluctance to get involved with details of administration. In such matters his adoption of Laozi's philosophy of "Wu Wei" (roughly translated as "Let Nature take its course") seems to have worked admirably.

Chern has always believed strongly that China could and should become a world leader in mathematics. But for this to happen he felt two preconditions were required:

(1) The existence within the Chinese mathematical community of a group of strong, confident, creative people, who are dedicated, unselfish, and aspire to go beyond their teachers, even as they wish their students to go beyond them.
(2) Ample support for excellent library facilities, research space, and communication with the world-wide mathematical community. (Chern claimed that these resources were as essential for mathematics as laboratories were for the experimental sciences).

It was to help in achieving these goals that Chern accepted the job of organizing the mathematics institute of Academia Sinica during 1946 to 1948, and the reason why he returned to Tianjin to found the Mathematics Institute at Nankai University after his retirement in 1984 as director of MSRI.

During 1965-76, because of the Cultural Revolution, China lost a whole generation of mathematicians, and with them much of the tradition of mathematical research. Chern started visiting China frequently after 1972, to lecture, to train Chinese mathematicians, and to rekindle these traditions. In part because of the strong bonds he had with Nankai University, he founded the Nankai Mathematical Research Institute there in 1985. This Institute has its own housing, and attracts many visitors both from China and abroad. In some ways it is modeled after the Institute for Advanced Study in Princeton. One of its purposes is to have a place where mature mathematicians and graduate students from all of China can spend a period of time in contact with each other and with foreign mathematicians, concentrating fully on research. Another is to have an inspiring place in which to work; one that will be an incentive for the very best young mathematicians who get their doctoral degrees abroad to return home to China.

Honors and awards

Chern was invited three times to address The International Congress of Mathematicians. He gave an Hour Address at the 1950 Congress in Cambridge, Massachusetts (the first ICM following the Second World War), spoke again in 1958, at Edinburgh, Scotland, and was invited to give a second Hour Address at the 1970 ICM in Nice, France. These Congresses are held only every fourth year and it is unusual for a mathematician to be invited twice to give a plenary Hour Address.

During his long career Chern was awarded numerous honorary degrees. He was elected to the US National Academy of Sciences in 1961, and received the National Medal of Science in 1975 and the Wolf Prize in 1983. The Wolf Prize was instituted in 1979 by the Wolf Foundation of Israel to honor scientists who had made outstanding contributions to their field of research. Chern donated the prize money he received from this award to the Nankai Mathematical Institute. He is also a foreign member of The Royal Society of London, Academie Lincei, and the French Academy of Sciences. A more complete list of the honors he received can be found in the Curriculum Vitae in [CSP].

An overview of Chern's research

Chern's mathematical interests have been unusually wide and far-ranging and he has made significant contributions to many areas of geometry, both classical and modern. Principal among these are:

- Geometric structures and their equivalence problems
- Integral geometry
- Euclidean differential geometry
- Minimal surfaces and minimal submanifolds
- Holomorphic maps
- Webs
- Exterior Differential Systems and Partial Differential Equations
- The Gauss-Bonnet Theorem
- Characteristic classes

Since it would be impossible within the space at our disposal to present a detailed review of Chern's achievements in so many areas, rather than attempting a superficial account of all facets of his research, we have elected to concentrate on those areas where the effects of his contributions have, in our opinion, been most profound and far-reaching. For further information concerning Chern's scientific contributions the reader may consult the four volume set, *Shiing-Shen Chern Selected Papers* [CSP]. This includes a Curriculum Vitae, a full bibliography of his published papers, articles of commentary by André Weil and Phillip Griffiths, and a scientific autobiography in which Chern comments briefly on many of his papers.

One further *caveat*; the reader should keep in mind that this is a mathematical biography, **not** a mathematical history. As such, it concentrates on giving an account of Chern's own scientific contributions, mentioning other mathematicians only if they were his coauthors or had some particularly direct and personal effect on Chern's research. Chern was working at the cutting edge of mathematics and there were of course many occasions when others made discoveries closely related to Chern's and at approximately the same time. A far longer (and different) article would have been required if we had even attempted to analyze such cases. But it is not only for reasons of space that we have avoided these issues. A full historical treatment covering this same ground would be an extremely valuable undertaking, and will no doubt one day be written. But that will require a major research effort of a kind that neither of the present authors has the training or qualifications even to attempt.

Before turning to a description of Chern's research, we would like to point out a unifying theme that runs through all of it: his absolute mastery of the techniques of differential forms and his artful application of these techniques in solving geometric problems. This was a magic mantle, handed down to him by his great teacher, Élie Cartan. It permitted him to explore in depth new mathematical territory where others could not enter. What makes differential forms such an ideal tool for studying local and global geometric properties (and for relating them to each other) is their two complementary aspects. They admit, on the one hand, the local operation of exterior differentiation, and on the other the global operation of integration over cochains, and these are related via Stokes' Theorem.

Geometric structures and their equivalence problems

Much of Chern's early work was concerned with various "equivalence problems". Basically, the question is how to determine effectively when two geometric structures of the same type are "equivalent" under an appropriate group of geometric transformations. For example, given two curves in space, when is there a Euclidean motion that carries one onto the other? Similarly, when are two Riemannian structures locally isometric? Classically one tried to associate with a given type of geometric structure various "invariants", that is, simpler and better understood objects that do not change under an isomorphism, and then show that certain of these invariants are a "complete set", in the sense that they determine the structure up to isomorphism. Ideally one should also be able to specify what values these invariants can assume by giving relations between them that are both necessary and sufficient for the existence of a structure with a given set of invariants. The goal is a theorem like the elegant classic paradigm of Euclidean plane geometry, stating that the three side lengths of a triangle determine it up to congruence, and that three positive real numbers arise as side lengths precisely when each is less than the sum of the other two. For smooth, regular space curves the solution to the equivalence problem was known early in the last century. If to a given space curve $\sigma(s)$ (parameterized by arc length) we associate its curvature $\kappa(s)$ and torsion $\tau(s)$, it is easy to show that these two smooth scalar functions are invariant under the group of Euclidean motions, and that they uniquely determine a curve up to an element of that group. Moreover any smooth real valued functions κ and τ can serve as curvature and torsion as long as κ is positive. The more complex equivalence problem for surfaces in space had also been solved by the mid 1800's. Here the invariants turned out to be two smooth quadratic forms on the surface, the first and second fundamental forms, of which the first, the metric tensor, had to be positive definite and the two had to satisfy the so-called Gauss and Codazzi equations. The so-called "form problem", that is the local equivalence problem for Riemannian metrics, was also solved classically (by Christoffel and Lipschitz). The solution is still more complex and superficially seems to have little in common with the other examples above.

As Chern was starting his research career, a major challenge facing geometry was to find what this seemingly disparate class of examples had in common, and thereby discover a general framework for the Equivalence Problem. Cartan saw this clearly, and had already made important steps in that direction with his general machinery of "moving frames". His approach was to reduce a general equivalence problem to one of a special class of equivalence problems for differential forms. More precisely, he would associate to a given type of local geometric structure in open sets U of $\mathbb{R}^n$, an "equivalent" structure, given by specifying:

1) a subgroup G of $\mathbf{GL}(n, \mathbb{R})$,
2) certain local co-frame fields $\{\theta_i\}$ in open subsets U of $\mathbb{R}^n$ (i.e., n linearly independent differential 1-forms in U).

The condition of equivalence for $\{\theta_i\}$ in U and $\{\theta_i^*\}$ in U^* is the existence of a diffeomorphism φ of U with U^* such that $\varphi^*(\theta_i^*) = \sum_{i=1}^{n} a_{ij}\theta_j$, where (a_{ij}) is a smooth map of U into G. A geometric structure defined by the choices 1) and 2) is now usually called a "G-structure", a name introduced by Chern in the course of formalizing and explicating Cartan's approach. For a given geometric structure one must choose the related G-structure so that its notion of equivalence coincides with that for the originally given geometric structure, so the invariants of the G-structure will also be the same as for the given geometric structure. In the case of the form problem one takes $G = O(n)$, and given a Riemannian metric ds^2 in U choose any θ_i such that $ds^2 = \sum_{i=1}^{n} \theta_i^2$ in U. While not always so obvious as in this case (and a real geometric insight is sometimes required for their discovery) most other natural geometric equivalence problems, including the ones mentioned above, do admit reformulation in terms of G-structures.

But do we gain anything besides uniformity from such a reformulation? In fact, we do, for Cartan also developed general techniques for finding complete sets of invariants for G-structures. Unfortunately, however, carrying out this solution of the Equivalence Problem in complete generality depends on his powerful but difficult theory of Pfaffian systems in involution, with its method of prolongation, a theory not widely known or well understood even today. In fact, while his preeminence as a geometer was clearly recognized towards the end of his career, many great mathematicians confessed to finding Cartan's work hard going at best, and few mathematicians of his day were able to comprehend fully his more novel and innovative advances. For example, in a review of one of his books (*Bull. Amer. Math. Soc.* vol. 44, p. 601) H. Weyl made this often quoted admission:

> "Cartan is undoubtedly the greatest living master in differential geometry...
> Nevertheless I must admit that I found the book, like most of Cartan's papers, hard reading..."

Given this well-known difficulty Cartan had in communicating his more esoteric ideas, one can easily imagine that his important insights on the Equivalence Problem might have lain buried. Fortunately they were spared such a fate.

Recall that Chern had spent his time at Hamburg studying the Cartan-Kähler theory of Pfaffian systems with Kähler, and immediately after Hamburg Chern spent a year in Paris continuing his study of these techniques with Cartan. Clearly Chern was ideally prepared to carry forward the attack on the Equivalence Problem. In a series of beautiful papers over the next twenty years not only did he do just that, but he also explained and reformulated the theory with such clarity and geometric appeal that much (though by no means all!) of the theory has become part of the common world-view of differential geometers, to be found in the standard textbooks on geometry. Those two decades were also, not coincidentally, the years that saw the development of the theory of fiber bundles and of connections on principal G-bundles. These theories were the result of the combined research efforts of many people and had multiple sources of inspiration both in topology and geometry. One major thread in that development was Chern's work on the Equivalence Problem and his related research on characteristic classes that grew out of it. In order to discuss this important work of Chern we must first define some of the concepts and notations that he and others introduced.

Using current geometric terminology, a G-structure for a smooth n-dimensional manifold M is a reduction of the structure group of its principal tangent co-frame bundle from $\mathbf{GL}(n, \mathbb{R})$ to the subgroup G. In particular, the total space of this reduction is a principal G-bundle, P, over M consisting of the admissible co-frames $\theta = (\theta_1, \dots, \theta_n)$, and we can identify the G-structure with this P. There are n canonically defined 1-forms ω_i on P; if $\Pi : P \to M$ is the bundle projection, then the value of ω_i at θ is $\Pi^*(\theta_i)$. The kernel of $D\Pi$ is of course the sub-bundle of the tangent bundle TP of P tangent to its fibers, and is usually called the vertical sub-bundle V. Clearly the canonical forms ω_i vanish on V. The group G acts on the right on P, acting simply transitively on each fiber, so we can identify the vertical space V_θ at any point θ with the Lie algebra $L(G)$ of left-invariant vector fields on G. Now, as Ehresmann first noted, a "connection" in Cartan's sense for the given G-structure (or as we now say, a G-connection for the principal bundle P) is the same as a "horizontal" sub-bundle H of TP complementary to V and invariant under G. Instead of H it is equivalent to consider the projection of TP onto V along H which, by the above identification of V_θ with $L(G)$, is an $L(G)$-valued 1-form ω on P, called the "connection 1-form". If we denote the right action of $g \in G$ on P by R_g, then the invariance of H under G translates to the transformation law $R_g^*(\omega) = ad(g^{-1}) \circ \omega$ for ω, where ad denotes the adjoint representation of G on $L(G)$. $L(G)$-valued forms on P transforming in this way are called *equivariant*. Since $L(G)$ is a sub-algebra of the Lie algebra $L(\mathbf{GL}(n, \mathbb{R}))$ of $n \times n$ matrices, we can regard

ω as an $n \times n$ matrix-valued 1-form on P, or equivalently as a matrix ω_{ij} of n^2 real-valued 1-forms on P.

If $\sigma : [0,1] \to M$ is a smooth path in M from p to q, then the connection defines a canonical G-equivariant map π_σ of the fiber P_p to the fiber P_q, called parallel translation along σ; namely $\pi_\sigma(\theta) = \tilde{\sigma}(1)$, where $\tilde{\sigma}$ is the unique horizontal lift of σ starting at θ. In general, parallel translation depends on the path σ, not just on the endpoints p and q. If it depends only on the homotopy class of σ with fixed endpoints, then the connection is called "flat". It is easy to see that this is so if and only if the horizontal sub-bundle H of TP is integrable, and using the Frobenius integrability criterion, this translates to $d\omega_{ij} = \sum_k \omega_{ik} \wedge \omega_{kj}$. Thus it is natural to define the matrix Ω_{ij} of so-called *curvature forms* of the connection, (whose vanishing is necessary and sufficient for flatness) by $d\omega_{ij} = \sum_k \omega_{ik} \wedge \omega_{kj} - \Omega_{ij}$ or, in matrix notation, $d\omega = \omega \wedge \omega - \Omega$. Since ω is equivariant, so is Ω. Differentiating the defining equation of the curvature forms gives the Bianchi identity, $d\Omega = \Omega \wedge \omega - \omega \wedge \Omega$. A local cross-section $\theta : U \to P$ is called an "admissible local co-frame" for the G-structure, and we can use it to pull back the connection forms and curvature forms to forms ψ_{ij} and Ψ_{ij} on U. Any other admissible co-frame field $\hat{\theta}$ in U is related to θ by a unique "change of gauge", g in U (i.e., a unique map $g : U \to G$) such that $\hat{\theta}(x) = R_{g(x)}\theta(x)$. If we use $\hat{\theta}$ to also pull back the connection and curvature forms to forms $\hat{\psi}$ and $\hat{\Psi}$ on U, then, using matrix notation, it follows easily from the equivariance of ω and Ω that $\hat{\psi} = dg\, g^{-1} + g\psi g^{-1}$ and $\hat{\Psi} = g\Psi g^{-1}$.

But where do connections fit into the Equivalence Problem? While Cartan's solution to the equivalence problem for G-structures was complicated in the general case, it became much simpler for the special case that G is the trivial subgroup $\{e\}$. For this reason Cartan had developed a method by which one could sometimes reduce a G-structure on a manifold M to an $\{e\}$-structure on a new manifold obtained by "adding variables" corresponding to coordinates in the group G. Chern recognized that this new manifold was just the total space P of the principal G-bundle, and that Cartan's reduction method amounted to finding an "intrinsic G-connection" for P, i.e., one canonically associated to the G-structure. Indeed the canonical 1-forms ω_i together with a linearly independent set of the connection forms ω_{ij}, defined by the intrinsic connection, give a canonical co-frame field for P, which of course is the same as an $\{e\}$-structure. Finally, Chern realized that in this setting one could describe geometrically the invariants for a G-structure given by Cartan's general method; in fact they can all be calculated from the curvature forms of the intrinsic connection.

Note that this covers one of the most important examples of a G-structure; namely the case $G = \mathbf{O}(n)$, corresponding to Riemannian geometry. The intrinsic connection is of course the "Levi-Civita connection". Moreover, in this case it is also easy to explain how to go on to "solve the form problem", i.e., to find explicitly a complete set of local invariants for a Riemannian metric. In fact, they can be taken as the components of the Riemann curvature tensor and its covariant derivatives in Riemannian normal coordinates. To see this, note first the obvious fact that there is a local isometry of the Riemannian manifold (M, g) with (M^*, g^*) carrying the orthonormal frame e_i at p to e_i^* at p^* if and only if in some neighborhood of the origin the components $g_{ij}(x)$ of the metric tensor of M with respect to the Riemannian normal coordinates x_k defined by e_i are **identical** to the corresponding components $g_{ij}^*(x)$ of the metric tensor of M^* with respect to Riemannian normal coordinates defined by e_i^*. The proof is then completed by using the easy, classical fact [ABP, Appendix II] that each coefficient in the Maclaurin expansion of $g_{ij}(x)$ can be expressed as a universal polynomial in the components of the Riemann tensor and a finite number of its covariant derivatives.

Let us denote by $N(G)$ the semi-direct product $G \ltimes \mathbb{R}^n$ of affine transformations of $\mathbb{R}^n$ generated by G and the translations. Correspondingly we can "extend" the principal G-bundle P of linear frames to the associated principal $N(G)$-bundle $N(P)$ of affine frames. Chern noted in [43] that the above technique could be expressed more naturally, and could be generalized to a wide class of groups G, if one looked for intrinsic $N(G)$-connections on $N(P)$. The curvature of an $N(G)$-connection on $N(P)$ is a two-form Ω on $N(P)$ with values in the Lie algebra $L(N(G))$ of $N(G)$. Now $L(N(G))$ splits canonically as the direct sum of $L(G)$ and $L(\mathbb{R}^n) = \mathbb{R}^n$, and Ω splits accordingly. The $\mathbb{R}^n$ valued part, τ, of Ω is called the torsion of the connection, and what Chern exploited was the fact that he could in certain cases define "intrinsic" $N(G)$ connections by putting conditions on τ. For example, the Levi-Civita connection can be characterized as the unique $N(\mathbf{O}(n))$ connection on $N(P)$ such that $\tau = 0$. In fact, in [43] Chern showed that if the Lie algebra $L(G)$ satisfied a certain simple algebraic condition ("property C") then it was always possible to define an intrinsic $N(G)$ connection in this way, and he proved that any compact G satisfies property C. He also pointed out here, from the point of view of Cartan's theory of pseudogroups, why some G-structures do **not** admit intrinsic connections. The pseudogroup of a G-structure $\Pi : P \to M$ is the pseudogroup of local diffeomorphisms of M whose differential preserves the subbundle P. It is elementary that the group of bundle automorphisms of a principal G-bundle that preserve a given G-connection is a finite dimensional Lie group and so *a fortiori* the pseudogroup of a G-structure with a canonically defined connection will be a Lie group. But there are important examples of groups G for which the pseudogroup of a G-structure is of infinite dimension. For example, if $n = 2m$ and we take $G = \mathbf{GL}(m, \mathbb{C})$, then a G-structure is the same thing as an almost-complex structure, and the group of automorphisms is an infinite pseudogroup.

Chern solved many concrete equivalence problems. In [6] and [13] he carried this out for the path geometry defined by a third order ordinary differential equation. Here the G-structure is on the contact manifold of unit tangent vectors of $\mathbb{R}^2$, and G is the ten-dimensional group of circle preserving contact transformations. In [10], [11] he generalized this to the path geometry of systems of n-th order ordinary differential equations. In [23] he considers a generalized projective geometry, i.e., the geometry of $(k+1)(n-k)$-parameter family of k-dimensional submanifolds in $\mathbb{R}^n$, and in [20], [21] the geometry defined by an $(n-1)$-parameter family of hypersurfaces in $\mathbb{R}^n$. In [105] (jointly with Moser) and in [106] he considers real hypersurfaces in $\mathbb{C}^n$. This latter research played a fundamental rôle in the development of the theory of CR manifolds.

Integral geometry

The group G of rigid motions of $\mathbb{R}^n$ acts transitively on various spaces S of geometric objects (e.g., points, lines, affine subspaces of a fixed dimension, spheres of a fixed radius) so that these spaces can be regarded as homogeneous spaces, G/H, and the invariant measure on G induces an invariant measure on S. This is the so-called "kinematic density", first introduced by Poincaré, and the basic problem of integral geometry is to express the integrals of various geometrically interesting quantities with respect to the kinematic density in terms of known integral invariants (see [84]). The simplest example is Crofton's formula for a plane curve C,

$$\int n(\ell \cap C)d\ell = 2L(C)$$

where $L(C)$ is its length, $n(\ell \cap C)$ is the number of its intersection points with a line ℓ in the plane, and $d\ell$ is the kinematic density on lines. We can interpret this formula as saying that the average number of times that a line meets a curve (i.e., is incident with a point on the curve) is equal to twice the length of the curve.

In [18], Chern laid down the foundations for a much generalized integral geometry. In [W], André Weil says of this paper that:

> "... it lifted the whole subject at one stroke to a higher plane than where Blaschke's school had left it, and I was impressed by the unusual talent and depth of understanding that shone through it."

Chern first extended the classical notion of "incidence" to a pair of elements from two homogeneous spaces G/H and G/K of the same group G. Given $aH \in G/H$ and $bK \in G/K$, Chern calls them "incident" if $aH \cap bK \neq \phi$. This definition plays an important rôle in the theory of Tits buildings.

In [48] and [84] Chern obtained fundamental kinematic formulas for two submanifolds in $\mathbb{R}^n$. The integral invariants in Chern's formula arise naturally in Weyl's formula for the volume of a tube T_ρ of radius ρ about a k-dimensional submanifold X of $\mathbb{R}^n$. Setting $m = n - k$, Weyl's formula is:

$$V(T_\rho) = \sum_{0 \leq i \leq k, i \text{ even}} c_i \mu_i(X) \rho^{m+i} .$$

Here the c_i are constants depending on m and i,

$$\mu_i(X) = \int_M I_i(\Omega)$$

where I_i is a certain adjoint invariant polynomial of degree $i/2$ on the Lie algebra of $\mathbf{O}(n)$, and Ω is the curvature form with respect to the induced metric on X. Chern's formula (also discovered independently by Federer) is:

$$\int \mu_e(M_1 \cap gM_2) dg = \sum_{0 \leq i \leq e, i \text{ even}} c_i \mu_i(M_1) \mu_{e-i}(M_2) ,$$

where M_1 and M_2 are submanifolds of $\mathbb{R}^n$ of dimensions p and q respectively, e is even, $0 \leq e \leq p + q - n$, and c_i are constant depend on n, p, q, e. Griffiths made the following comment concerning this paper [G]:

"Chern's proof of [this formula] exhibits a number of characteristic features. Of course, one is the use of moving frames.... Another is that the proof proceeds by direct computation rather than by establishing an elaborate, conceptual framework; in fact upon closer inspection there is such a conceptual framework, as described in [18], however, the philosophical basis is not isolated but is left to the reader to understand by seeing how it operates in a non-trivial problem."

Euclidean differential geometry

One of the main topics in classical differential geometry is the study of local invariants of submanifolds in Euclidean space under the group of rigid motions, i.e., the equivalence problem for submanifolds. The solution is classical. In fact, the first and second fundamental forms, I and II, and the induced connection ∇^ν on the normal bundle of a submanifold satisfy the Gauss, Codazzi and Ricci equations, and they form a complete set of local invariants for submanifolds in $\mathbb{R}^n$. Explicitly these invariants are as follows:

a) I is the induced metric on M,

b) II is a quadratic form on M with values in the normal bundle $\nu(M)$ such that, for any unit tangent vector u and unit normal vector v at p, $II_v(u) = \langle II(u), v\rangle$ is the curvature at p of the plane curve σ formed by intersecting M with the plane spanned by u and v, and

c) if s is a smooth normal field then $\nabla^\nu(s)$ is the orthogonal projection of the differential ds onto the normal bundle $\nu(M)$.

$II_\nu = \langle II, \nu\rangle$ is called the second fundamental form in the direction of ν. The self-adjoint operator A_ν corresponding to II_ν is called the shape operator of M in the direction ν.

Chern's work in this field involved mainly the relation between the global geometry of submanifolds and these local invariants. He wrote many important papers in the area, but because of space limitations we will concentrate only on the following:

(1) Minimal surfaces

Since the first variation for the area functional for submanifolds of $\mathbb{R}^n$ is the trace of the second fundamental form, a submanifold M of $\mathbb{R}^n$ is called *minimal* if $\mathrm{trace}(II) = 0$. Let $\mathbf{Gr}(2,n)$ denote the Grassmann manifold of 2-planes in $\mathbb{R}^n$. The Gauss map G of a surface M in $\mathbb{R}^n$ is the map from M to the Grassmann manifold $\mathbf{Gr}(2,n)$ defined by $G(x)$ = the tangent plane to M at x. The Grassmann manifold $\mathbf{Gr}(2,n)$ can be identified as the hyperquadric $z_1^2 + \cdots + z_n^2 = 0$ of $\mathbb{C}\mathbf{P}^{n-1}$ (via the map that sends a 2-plane V of $\mathbb{R}^n$ to the complex line spanned by $e_1 + ie_2$, where (e_1, e_2) is an orthonormal base for V). Thus $\mathbf{Gr}(2,n)$ has a complex structure. On the other hand, an oriented surface in $\mathbb{R}^n$ has a conformal and hence complex structure through its induced Riemannian metric. In [79], Chern proved that an immersed surface in $\mathbb{R}^n$ is minimal if and only if the Gauss map is antiholomorphic. This theorem was proved by Pinl for $n = 4$ and is the starting point for relating minimal surfaces with the value distribution theory of Nevanlinna, Weyl, and Ahlfors. One of the fundamental results of minimal surface theory is the Bernstein uniqueness theorem, which says that a minimal graph $z = f(x,y)$ in $\mathbb{R}^3$, defined for all $(x,y) \in \mathbb{R}^2$, must be a plane. Note that the image of the Gauss map of an entire graph lies in a hemisphere. Bernstein's theorem as generalized by Osserman says that if the image of the Gauss map of a complete minimal surface of $\mathbb{R}^3$ is not dense, then the minimal surface is a plane. In [79], using a classical theorem of E. Borel, Chern generalized the Bernstein-Osserman theorem to a density theorem on the image of the Gauss map of a complete minimal surface in $\mathbb{R}^n$, that is not a plane. More refined density theorems were established in [86], a joint paper with Osserman.

Motivated by Calabi's work on minimal 2-spheres in $\mathbb{S}^n$, Chern developed in [96] a general formalism for osculating spaces for submanifolds. He proved that given a minimal surface in a space form there is an integer m such that the osculating spaces of order m are parallel along the surface, and gave a complete system of local invariants, with their relations. As a consequence, he proved an analogue of Calabi's theorem: if a minimal sphere of constant Gaussian curvature K in a space form of constant sectional curvature c is not totally geodesic, then $K = 2c/m(m+1)$.

(2) Tight and taut immersions

We first recall a theorem of Fenchel, proved in 1929: if $\alpha(s)$ is a simple closed curve in $\mathbb{R}^3$, parametrized by its arc length, and $k(s)$ is its curvature function, then $\int |k(s)|ds \geq 2\pi$, and equality holds if and only if α is a convex plane curve. Fary and Milnor proved that if α is knotted then this integral must be greater than 4π.

In [62] and [66], Chern and Lashof generalized these results to submanifolds of $\mathbb{R}^n$. Let M^m be a compact m-dimensional manifold, $f : M \to \mathbb{R}^n$ an immersion, $\nu^1(M)$ the unit normal sphere bundle of M, and dv the natural volume element of $\nu^1(M)$. Let $N : \nu^1(M) \to \mathbb{S}^{n-1}$ denote the normal map, i.e., N maps the unit normal vector v at x to the parallel unit vector at the origin. Let da denote the volume element of S^{n-1}. Then the Lipschitz-Killing curvature G on $\nu^1(M)$ is defined by the equation $N^*(da) = GdA$, i.e., $G(v)$ is the absolute value of the determinant of the shape operator A_v of M along the unit normal direction v. The absolute total curvature $\tau(M, f)$ of the immersion f is the normalized volume of the image of N,

$$\tau(M, f) = \frac{1}{c_{n-1}} \int_{\nu^1(M)} |\det(A_v)| dv ,$$

where c_{n-1} is the volume of the unit $(n-1)$-sphere.

In [62] Chern and Lashof generalized Fenchel's theorem by showing that $\tau(M, f) \geq 2$, with equality if and only if M is a convex hypersurface of an $(m+1)$-dimensional affine subspace V. In [66] they obtained the sharper result that

$$\tau(M,f) \geq \sum \beta_i(M) ,$$

where $\beta_i(M)$ is the i-th Betti number of M.

An immersion $f : M \to \mathbb{R}^n$ is called *tight* if $\tau(M, f)$ is equal to the infimum, $\tau(M)$, of the absolute total curvature among all the immersions of M into Euclidean spaces of arbitrary dimensions. The study of absolute total curvature and tight immersion has become an important field in submanifold geometry that has seen many interesting developments in recent years. An important step in this development is Kuiper's reformulation of tightness in terms of critical point theory. He showed that for a given compact manifold M, $\tau(M)$ is the Morse number γ of M, i.e., the minimum number of critical points a non-degenerate Morse function must have. Moreover, an immersion of M is tight if and only if every non-degenerate height function has exactly $\tau(M) = \gamma$ critical points. Another development is the concept of taut immersion introduced by Banchoff and Carter-West. An immersion of M into $\mathbb{R}^n$ is called *taut* if every non-degenerate Euclidean distance function from a fixed point in $\mathbb{R}^n$ to the submanifold has exactly γ critical points. Taut implies tight, and moreover a taut immersion is an embedding. Tautness is invariant under conformal transformations, hence using stereographic projection we may assume taut submanifolds lie in the sphere. Pinkall proved that the tube M_ϵ of radius ϵ around a submanifold M in $\mathbb{R}^n$ is a taut hypersurface if and only if M is a taut submanifold. In particular, this gives two facts: one is that the parallel hypersurface of a taut hypersurface in $\mathbb{S}^n$ is again taut, another is that to understand taut submanifolds it suffices to understand taut hypersurfaces. Since the Lie sphere group (the group of contact transformations carrying spheres to spheres) is generated by conformal transformations and parallel translations, tautness is invariant under the Lie sphere group. Note also that the ϵ-tube M_ϵ of a submanifold

M in S^n is an immersed Legendre submanifold of the contact manifold of the unit tangent bundle of S^n. Thus tautness really should be defined for Legendre submanifolds of the contact manifold of the unit tangent sphere bundle of $\mathbb{S}^n$. Chern and Cecil make this concept precise in [143] and lay some of the basic differential geometric groundwork for Lie sphere geometry. There are many interesting examples of tight and taut submanifolds and many interesting theorems concerning them. But some of the most basic questions are still unanswered; for example there are no good necessary and sufficient conditions known for a compact manifold to be immersed in Euclidean space as a tight or taut submanifold, and a complete set of local invariants for Lie sphere geometry is yet to be found.

The Generalized Gauss-Bonnet Theorem

Geometers tend to make a sharp distinction between "local" and "global" questions, and it is common not only to regard global problems as somehow more important, but even to consider local theory "old-fashioned" and unworthy of serious effort. Chern however has always maintained that research on these seemingly polar aspects of geometry must of necessity go hand-in-hand; he felt that one could not hope to attack the global theory of a geometric structure until one understood its local theory (i.e., the equivalence problem), and moreover, once one had discovered the local invariants of a theory, one was well on the way towards finding its global invariants as well! We shall next explain how Chern came to this contrary attitude, for it is an interesting and revealing story, involving the most exciting and important events of his research career: his discovery of an "intrinsic proof" of the Generalized Gauss-Bonnet Theorem and, flowing out of that, his solution of the characteristic class problem for complex vector bundles by his striking and elegant construction of what are now called "Chern classes" from his favorite raw material, the curvature forms of a connection.

The Gauss-Bonnet Theorem for a closed, two-dimensional Riemannian manifold M was surely one of the high points of classical geometry, and it was generally recognized that generalizing it to higher dimensional Riemannian manifolds was a central problem of global differential geometry. The theorem states that the most basic topological invariant of M, its Euler characteristic $\chi(M)$, can be expressed as $1/2\pi$ times the integral over M of its most basic geometric invariant, the Gaussian curvature function K. Although there were many published proofs of this, Chern reproved it for himself by a new method that was very natural from a moving frames perspective. Moreover, unlike the published proofs, Chern's had the potential to generalize to higher dimensions.

To explain Chern's method, we start by applying the standard moving frames approach to n-dimensional oriented Riemannian manifolds M, then specialize to $n = 2$. The orientation together with the Riemannian structure give an $\mathbf{SO}(n)$ structure for M. Since the Lie algebra $L(\mathbf{SO}(n))$ is just the skew-adjoint $n \times n$ matrices, in the principal $\mathbf{SO}(n)$ bundle $F(M)$ of oriented orthonormal frames of M, in addition to the n canonical 1-forms (ω_i), we will have the connection 1-forms for the Levi-Civita connection, a skew-adjoint $n \times n$ matrix of 1-forms ω_{ij}, characterized uniquely by the zero torsion condition, $d\omega_i = \sum_j \omega_{ij} \wedge \omega_j$. The components R_{ijkl} of the Riemann curvature tensor in the frame ω_i are determined from the curvature forms Ω_{ij} by $\Omega_{ij} = \frac{1}{2} \sum_{kl} R_{ijkl}\omega_k \wedge \omega_l$ (plus the condition of being skew-symmetric in (i, j) and in (k, l)).

When $n = 2$, the Lie algebra $L(\mathbf{SO}(n))$ is 1-dimensional; $\omega_{11} = \omega_{22} = 0$ and $\omega_{21} = -\omega_{12}$, so there is only one independent ω_{ij}, namely ω_{12}, and so only one curvature equation, $d\omega_{12} = -\Omega_{12} = -R_{1212}\omega_1 \wedge \omega_2$. Now it is easily seen that R_{1212} is a constant on every fiber $\Pi^{-1}(x)$, and its value is in fact the Gaussian curvature $K(x)$. We can identify the area 2-form, dA, on M with $-\theta_1 \wedge \theta_2$, where (θ_1, θ_2) is any oriented orthonormal frame, so that $\Pi^*(dA) = -\omega_1 \wedge \omega_2$. Thus we can rewrite the above curvature equation as a formula for the pull-back of the Gauss-Bonnet integrand, $K\ dA$, to $F(M)$:

$$\Pi^*(K\ dA) = d\omega_{12}\ . \qquad (*)$$

In [136] Chern remarks that, along with zero torsion equations, the formula (∗) contains

> "... all the information on local Riemannian geometry in two dimensions [and] gives global consequences as well. A little meditation convinces one that (∗) must be the formal basis of the Gauss-Bonnet formula, and this is indeed the case. It turns out that the proof of the n-dimensional Gauss-Bonnet formula can based on this idea ..."

Chern noticed a remarkable property of (∗). Since the Gauss-Bonnet integrand is a 2-form on a 2-dimensional manifold, it is automatically closed, and hence its pull-back under Π^* must also be closed. But (except when M is a torus) $K\ dA$ is never exact, so we do not expect its pull back to be exact. Nevertheless, (∗) says that it is! This phenomenon of a closed but non-exact form on the base of a fiber bundle becoming exact when pulled up to the total space is called *transgression*. As we shall see, it plays a key rôle in Chern's proof.

By elementary topology, in the complement M' of any point p of a closed Riemannian manifold M one can always define a smooth vector field e_1 of unit length, and the index of this vector field at p is $\chi(M)$. We will now see how this well-known characterization of the Euler characteristic together with the transgression formula (∗) leads quickly to Chern's proof of the Gauss-Bonnet theorem for two-dimensional M. Let e_2 denote the unit length vector field in M' making (e_1, e_2) an oriented frame, and let θ denote the dual co-frame field in M'. Since Π composed with θ is the identity map of M', we have $d(\theta^*(\omega_{12})) =$

$\theta^*(d\omega_{12}) = K\ dA$ in M', so $\int_M K\ dA = \int_{M'} K\ dA = \int_{M'} d(\theta^*(\omega_{12}))$. If we write M_ϵ for the complement of the open ϵ-ball about p, then $\int_{M'} = \lim_{\epsilon \to 0} \int_{M_\epsilon}$, and by Stokes' Theorem, $\int_M K\ dA = \lim_{\epsilon \to 0} \int_{S_\epsilon} \theta^*(\omega_{12})$, where $S_\epsilon = \partial M_\epsilon$ is the distance sphere of radius ϵ about p. The proof will be complete if we can identify the right hand side of the latter equation with 2π times the index of e_1 at p.

Choose Riemannian normal coordinates in a neighborhood U of p and let $(\hat{e}_1, \hat{e}_2)$ denote the local frame field in U defined by orthonormalizing the corresponding coordinate basis vectors, and $\hat{\theta}$ the dual co-frame field. If $\alpha(x)$ denotes the angle between $e_1(x)$ and $\hat{e}_1(x)$, then we recall that the standard expression for the index or winding number of e_1 with respect to p is $\frac{1}{2\pi} \int_C d\alpha$ where C is a small simple closed curve surrounding p; so we will be done if we can show that the right hand side above is equal to $\int_{S_\epsilon} d\alpha$.

Let $\rho(\alpha) \in \mathbf{SO}(2)$ denote rotation through an angle α. The gauge transformation $g : U \to \mathbf{SO}(2)$ from the co-frame $\hat{\theta}$ to the co-frame θ is just $g(x) = \rho(\alpha(x))$, so by the transformation law for pull-backs of connection forms noted above, $\theta^*(\omega_{12}) = d\alpha + \hat{\theta}^*(\omega_{12})$. Thus $\int_{S_\epsilon} \theta^*(\omega_{12})$ can be written as the sum of two terms. The first is the desired $\int_{S_\epsilon} d\alpha$, and the second term, $\int_{S_\epsilon} \hat{\theta}^*(\omega_{12})$ clearly tends to zero with ϵ since the integrand is continuous at p, while the length of S_ϵ tends to zero.

We now return to the case of a general n-dimensional oriented Riemannian manifold M and develop some machinery we will need to explain the remarkable results that grew out of this approach to the two-dimensional Gauss-Bonnet Theorem.

A basic problem is how to construct differential forms on M canonically form the metric. Up in the co-frame bundle, $F(M)$, there is an easy way to construct differential forms naturally from the metric–simply take "polynomials" in the curvature forms Ω_{ij}. Certain forms Λ constructed this way will "define" a form λ on M by the relation $\Lambda = \Pi^* \lambda$, and these are the forms we are after.

To make this precise we consider the ring $\mathcal{R}$ of polynomials with real (or complex) coefficients in $n(n-1)/2$ variables $\{X_{ij}\}$, $1 \leq i < j \leq n$. We use matrix notation; X denotes the $n \times n$ matrix X_{ij} of elements of $\mathcal{R}$, where $X_{ji} = -X_{ij}$ for $i < j$, and $X_{ii} = 0$. For $g \in \mathbf{SO}(n)$, $ad(g)X = gXg^{-1}$ is the matrix $\sum_{k,l} g_{ik}X_{kl}g_{jl}$ of elements of $\mathcal{R}$. If for g in $\mathbf{SO}(n)$ and P in $\mathcal{R}$ we define $\mathrm{ad}(g)P$ in $\mathcal{R}$ by $(\mathrm{ad}(g)P)(X) = P(\mathrm{ad}(g)X)$, this defines an "adjoint" action of $\mathbf{SO}(n)$ on $\mathcal{R}$ (by ring automorphisms). The subring of "ad-invariant" elements of $\mathcal{R}$ is denoted by $\mathcal{R}^{\mathrm{ad}}$. For future reference we note that we can also regard X as representing the general $n \times n$ skew-symmetric matrix, i.e., the general element of the Lie algebra $L(\mathbf{SO}(n))$, and $\mathcal{R}$ is just the ring of polynomial functions on $L(\mathbf{SO}(n))$.

The curvature 2-forms Ω_{ij}, being of even degree, commute with each other under exterior multiplication, so we can substitute them in elements P of $\mathcal{R}$; if $P(X)$ is homogeneous of degree d in the X_{ij}, then $P(\Omega)$ will be a differential $2d$-form on $F(M)$.

Now let θ be a local orthonormal co-frame field in an open set U of M, i.e., a local section $\theta : U \to F(M)$, and let $\Psi = \theta^*(\Omega)$ denote the matrix of pulled back curvature forms in U. Since θ^* is a Grassmann algebra homomorphism, for any P in $\mathcal{R}$, $\theta^*(P(\Omega)) = P(\Psi)$. In particular for any x in U we have $\theta^*(P(\Omega))_x = P(\Psi_x)$. If $\hat{\Psi}$ is the matrix of curvature forms in U corresponding to some other local co-frame field, $\hat{\theta}$ in U, and $g : U \to \mathbf{SO}(n)$ is the change of gauge mapping θ to $\hat{\theta}$, then as noted above, $\hat{\Psi}_x = \mathrm{ad}(g(x))\Psi$, so we find $P(\hat{\Psi}_x) = (\mathrm{ad}(g(x))P)(\Psi_x)$. Thus in general the pulled back form $P(\Psi)$ depends on the choice of θ and is only defined locally, in U. However *if (and only if) P is in the subring $\mathcal{R}^{\mathrm{ad}}$ of* ad*-invariant polynomials, the form $P(\Psi)$ is a globally well-defined form on M, independent of the choice of local frame fields θ used to pull back the locally-defined curvature matrices Ψ.* In this case it is clear that $\Pi^*(P(\Psi)) = P(\Omega)$, a relation that uniquely determines $P(\Psi)$.

There are many ways one might attempt to generalize the Gauss-Bonnet Theorem for surfaces, but perhaps the most obvious and natural is to associate with every compact, oriented, n-dimensional Riemannian manifold without boundary, M, an n-form λ on M that is canonically defined from the metric, and has the property that $\int_M \lambda = c_n \chi(M)$, where c_n is some universal constant. If n is odd then Poincaré duality implies that $\chi(M) = 0$ when M is without boundary, and since we will only consider the closed case here, we will assume $n = 2k$. (On the other hand, for odd-dimensional manifolds **with** boundary, the Gauss-Bonnet Theorem is interesting and decidedly non-trivial!). From the above discussion it is clear that we should define $\lambda = P(\Psi)$, where P is an ad-invariant polynomial, homogeneous of degree k in the X_{ij}. In fact there is an obvious candidate for P–the classical Pfaffian, Pf, uniquely determined (up to sign) by the condition that $\mathrm{Pf}(X)^2 = \det(X)$ (cf. [MS], page 309).

A Generalized Gauss-Bonnet Theorem had already been proved in two papers, one by Allendoerfer and the other by Fenchel. Both proofs were "extrinsic"–they assumed M could be isometrically embedded in some Euclidean space. (A paper of Allendoerfer and Weil implied that the existence of local isometric embeddings was enough, thereby settling the case of analytic metrics). These earlier proofs wrote the Generalized Gauss-Bonnet integrand as the volume element times a scalar that was a complicated polynomial in the components of the Riemann tensor. In [25] Chern for the first time wrote the integrand as the Pfaffan of the curvature forms and then provided a simple and elegant **intrinsic** proof of the theorem along the lines of the above proof for surfaces.

Let $\mathbf{S}(M)$ denote the bundle of unit vectors of the tangent bundle to M, and $\pi : \mathbf{S}(M) \to M$ the natural projection. Given a co-frame θ in $F(M)$ let $e_1(\theta)$ denote the first element of the frame dual to θ. Then $e_1 : F(M) \to \mathbf{S}(M)$ is a fiber bundle and clearly $\Pi : F(M) \to M$ factors as $\Pi = \pi \circ e_1$. Let λ be the n-form Pf(Ψ) on M, and $\Lambda = p^*(\lambda)$ its pull-back to $\mathbf{S}(M)$. In [25] Chern first proves a transgression lemma for Λ, i.e, he explicitly finds an $n-1$-form Θ on $\mathbf{S}(M)$ satisfying $d\Theta = \Lambda$. As in two dimensions let M' be the complement of some point p in M and construct a smooth cross-section ξ of $\mathbf{S}(M)$ over M'. Then $\pi \circ \xi$ is the identity map of M', so just as in the two dimensional argument we find $d(\xi^*(\Theta)) = \lambda$, and $\int_M \lambda = \lim_{\epsilon \to 0} \int_{S_\epsilon} \xi^*(\Theta)$. Finally, the construction of Θ is so explicit that Chern is able to evaluate the right hand side by an argument similar to the one in the surface case, and he finds that it is indeed a universal constant times the Euler characteristic of M.

Mathematicians in general value proofs of new facts much more highly than elegant new proofs of old results. It is worth commenting why [25] is an exception to this rule. The earlier proofs of the Generalized Gauss-Bonnet Theorem were virtually a dead end while, as we shall see below, Chern's intrinsic proof was a key that opened the door to the secrets of characteristic classes.

Characteristic classes

The co-frame bundle, $F(M)$, that keeps re-appearing in our story, is an important example of a mathematical structure known as a *principal G-bundle*. These were first defined and their study begun only in the late 1930's, but their importance was quickly recognized by topologists and geometers, and the theory underwent intensive development during the 1940's. By the end of that decade the beautiful classification theory had been worked out, and with it the related theory of "characteristic classes", a concept whose importance for the mathematics of the latter half of the twentieth century it would be difficult to exaggerate. (As we will see below, in the language we have been using, the classification problem is the equivalence problem for principal bundles, and characteristic classes are invariants for this equivalence problem).

In order to explain Chern's rôle in these important developments we will first review some of the basic mathematical background of the theory.

We will consider only the case of a Lie group G. Since the theory is essentially the same for a Lie group and one of its maximal compact subgroups, we will also assume that G is compact. A "space" will mean a paracompact topological space, and a G-space will mean a space, P, together with a continuous right action of G on P. We will write R_g for the homeomorphism $p \mapsto pg$. The G-space P is called a *principal G-bundle* if the action is free, i.e., if for all p in P, $R_g(p) \neq p$ unless g is the identity element e of G. More specifically, P is called a principal G-bundle over a space X if we are given some fixed homeomorphism of X with the orbit space P/G, or equivalently if there is given a "projection map" $\Pi : P \to X$ such that the G orbits of P are exactly the "fibers" $\Pi^{-1}(x)$ of the map Π. P is called the *total space* of the bundle, and we often denote the bundle by the same symbol as the total space. A map $\sigma : X \to P$ that is a left inverse to Π is called a section. Two G-bundles over X, $\Pi_i : P_i \to X$, $i = 1, 2$ are considered "equivalent" if there is a G-equivariant homeomorphism $\varphi : P_1 \to P_2$ such that $\Pi_1 = \Pi_2 \circ \varphi$. The principal G-bundle over X defined by $P = X \times G$ with $R_g(x, \gamma) = (x, \gamma g)$ and $\Pi(x, \gamma) = x$ is called the product bundle, and any bundle equivalent to the product bundle is called a trivial bundle. Clearly $x \mapsto (x, e)$ is a section of the product bundle, so any trivial bundle has a section. Conversely, if $\Pi : P \to X$ has a section σ, then $\varphi(x, g) = R_g(\sigma(x))$ is an equivalence of the product bundle with P, i.e., *a principal G-bundle is trivial if and only if it admits a section.* We will denote the set of equivalence classes $[P]$ of principal G-bundles P over X by $\mathrm{Bndl}_G(X)$.

Given a principal G-bundle $\Pi : P \to X$ and a continuous map $f : Y \to X$, we can define a bundle $f^*(P)$ over Y, called the bundle induced from P by the map f. Its total space is $\{(p, y) \in P \times Y | \Pi(p) = f(y)\}$, with the projection $(p, y) \mapsto y$ and the G-action $R_g(p, y) = (R_g(p), y)$. It is easy to see that f^* maps equivalent bundles to equivalent bundles, so it induces a map (also denoted by f^*) from $\mathrm{Bndl}_G(X)$ to $\mathrm{Bndl}_G(Y)$. If $\Pi : P \to X$ is a principal G-bundle then $\Pi^*(P)$ is a principal G-bundle over the total space P, called the "square" of the original bundle. *In fact this bundle is always trivial*, since it admits the "diagonal" section $p \mapsto (p, p)$. As we will see below, this simple observation is the secret behind transgression!

The first non-trivial fact in the theory is the so-called "covering homotopy theorem"; it says that the induced map $f^* : \mathrm{Bndl}_G(X) \to \mathrm{Bndl}_G(Y)$ depends only on the homotopy class $[f]$ of f. We can paraphrase this by saying that $\mathrm{Bndl}_G(\ )$ is a contravariant functor from the category of spaces and homotopy classes of maps to the categroy of sets. Now a cohomology theory is also such a functor, and a *characteristic class* for G-bundles can be defined as simply a natural transformation from $\mathrm{Bndl}_G(\ )$ to some cohomology theory $H^*(\ )$. Of course this fancy language isn't essential and was only invented about the same time as bundle theory. It just says that a characteristic class c is a function that assigns to each principal G-bundle P over any space X an element $c(P)$ in $H^*(X)$, with the "naturality" property that $c(f^*(P)) = f^*(c(P))$, for any continuous $f : Y \to X$. We fix some cohomology theory $H^*(\ )$ and denote by $\mathrm{Char}(G)$ the set of all characteristic classes for G-bundles. Since $H^*(X)$ has the structure of a ring with unit, so does $\mathrm{Char}(G)$, and the characteristic class problem for G is the problem of explicitly identifying this ring. Note that a trivial bundle is induced from a map to a space with one point, so all its characteristic classes (except the unit class) must be zero. More generally, equality of all characteristic classes of a bundle is a necessary (and in some circumstances sufficient) test for their equivalence, and this is one of the important uses of characteristic classes.

The remarkable and beautiful classification theorem for principal

G-bundles "solves" the classification problem at least in the sense of reducing it to a standard problem of homotopy theory. Given spaces X and Z let $[X, Z]$ denote the set of homotopy classes of maps of X into Z. Note that $[\ , Z]$ is a contravariant functor, much like Bndl_G– any map $f : Y \to X$ induces a pull-back map $f^* : [h] \mapsto [h \circ f]$ of $[X, Z]$ to $[Y, Z]$. Moreover if $\Pi : P \to Z$ is any principle G-bundle then we have a map $[h] \mapsto [h^*(P)]$ of $[X, Z]$ to $\mathrm{Bndl}_G(X)$ that is "natural" (i.e., it commutes with all "pull-back" maps f^*). We call P a *universal* principal G-bundle if the latter map is bijective. *The heart of the classification theorem is the fact that universal G-bundles do exist.* In fact it can be shown that a principal G-bundle is universal provided its total space is contractible, and there are even a number of methods for explicitly constructing such bundles.

We will denote by $\mathcal{U}_G$ some choice of universal principal G-bundle. Its base space will be denoted by $\mathcal{B}_G$ and is called the *classifying space* for G. (Although $\mathcal{B}_G$ is not unique, its homotopy type is). If $\Pi : P \to X$ is any principal G-bundle then, by definition of universal, there is a **unique** homotopy class $[h]$ of maps of X to $\mathcal{B}_G$ such that P is equivalent to $h^*(\mathcal{U}_G)$. Any representative h is called a *classifying map* for P. Clearly if $f : Y \to X$ then $h \circ f$ is a classifying map for $f^*(P)$. Also, the classifying map for $\mathcal{U}_G$ is just the identity map of $\mathcal{B}_G$.

It is now easy to give a solution of sorts to the characteristic class problem for G; namely $\mathrm{Char}(G)$ *is canonically isomorphic to* $H^*(\mathcal{B}_G)$. In fact each $c \in H^*(\mathcal{B}_G)$ defines a characteristic class (also denoted by c) by the forumula $c(P) = f^*(c)$, where f is a classifying map for P, and the inverse map is just $c \mapsto c(\mathcal{U}_G)$.

This is a distillation of ideas developed between 1935 and 1950 by Chern, Ehresmann, Hopf, Feldbau, Pontrjagin, Steenrod, Stiefel, and Whitney. While elegant in its simplicity, the above version is still too abstract and general to be of use in finding $\mathrm{Char}(G)$ for a specific group G. It is also of little use in calculating the characteristic classes of bundles that come up in geometric problems, for it is not often an easy matter to find a classifying map from geometric data. We shall discuss how Chern put flesh on these bones by finding concrete models for classifying spaces and, more importantly, by showing how to calculate explicitly de Rham theory representatives of many characteristic classes from the curvature forms of connections.

Let $\mathbf{V}(n, N+n)$ denote the *Stiefel manifold* of n-frames in $\mathbb{R}^{N+n}$, consisting of all orthonormal squences $e = (e_1, \dots, e_n)$ of vectors in $\mathbb{R}^{N+n}$. There is an obvious free action of $\mathbf{O}(n)$ on $\mathbf{V}(n, N+n)$, and the orbit of e consists of all n-frames spanning the same n-dimensional linear subspace that e does. Thus we have an $\mathbf{O}(n)$ principal bundle $\Pi : \mathbf{V}(n, N+n) \to \mathbf{Gr}(n, N+n)$, where $\mathbf{Gr}(n, N+n)$ is the Grassmannian of all n-dimensional linear subspaces of $\mathbb{R}^{N+n}$. In the early 1940's it was known from results of Steenrod and Whitney that this bundle is "universal for compact k-dimensional polyhedra", provided $N \geq k+1$. This means that for any compact polyhedral space X, with $\dim(X) \leq k$, every principal $\mathbf{O}(n)$ bundle over X is of the form $h^*(\mathbf{V}(n, N+n))$ for a unique $[h]$ in $[X, \mathbf{Gr}(n, N+n)]$. In [43] Chern and Y.F. Sun generalized these results to show that this bundle is also universal for compact k-dimensional ANR's. (If one wants universal bundles in the strict sense described above, one need only form the obvious inductive limit, $\Pi : \mathbf{V}(n, \infty) \to \mathbf{Gr}(n, \infty)$, by letting N tend to infinity. But for the finite dimensional problems of geometry it is preferable to stick with these finite dimensional models). By replacing the real numbers respectively by the complex numbers and the quaternions, Chern and Sun proved analogous results for the other classical groups $\mathbf{U}(n)$ and $\mathbf{Sp}(n)$. They went on to note that if G is any compact Lie group, then by taking a faithful representation of G in some $\mathbf{O}(n)$, $\mathbf{V}(n, N+n)$ becomes a principal G bundle by restriction, and the corresponding orbit space $\mathbf{V}(n, N+n)/G$ becomes a classifying space $\mathcal{B}_G$ for compact ANR's of dimension $\leq k$.

The Grassmannians make good models for classifying spaces, for they are well-studied explicit objects whose cohomology can be investigated using both algebraic and geometric techniques. From such computations Chern knew that there was an n-dimensional "Euler class" e in Char($\mathbf{SO}(n)$). If M is a smooth, compact, oriented n-dimensional manifold then $e(F(M)) \in H^n(M)$ when evaluated on the fundamental class of M is just $\chi(M)$. One can thus interpret the Generalized Gauss-Bonnet Theorem as saying that $\lambda = \mathrm{Pf}(\Psi)$ represents $e(F(M))$ in de Rham cohomology. This inspired Chern to look for a general technique for representing characteristic classes by de Rham classes. This was in 1944-1945, while Chern was in Princeton, and he discussed this problem frequently with his friend André Weil who encouraged him in this search.

It might seem natural to start by trying to represent $\mathbf{SO}(n)$ characteristic classes by closed differential forms, but Chern made what was to be a crucial observation: the cohomology of the real Grassmannians is complicated. In particular it contains a lot of $\mathbb{Z}_2$ torsion, and this part of the cohomology is invisible to de Rham theory. On the other hand Chern knew that Ehresmann, in his thesis, had calculated the homology of complex Grassmannians and showed there was no torsion. In fact Ehresmann showed that certain explicit algebraic cycles (the "Schubert cells") form a free basis for the homology over $\mathbb{Z}$. It follows from de Rham's Theorem that all the cohomology classes for $\mathcal{B}_{U(n)}$ can be represented by closed differential forms. These forms, when pulled back by the classifying map of a principal $\mathbf{U}(n)$-bundle, will then represent the characteristic classes of the bundle in de Rham cohomology. While this is fine in theory, it still depends on knowing a classifying map, while what is needed in practice is a method to calculate these characteristic forms from geometric data. We now explain Chern's beautiful algorithm for doing this.

Let $\Pi : P \to M$ be a smooth principle $\mathbf{U}(n)$-bundle over a smooth manifold M. Recall that a connection for P can be regarded as a 1-form ω on P with values in the Lie algebra of $\mathbf{U}(n)$, $L(\mathbf{U}(n))$, which consists of all $n \times n$ skew-hermitian complex matrices. Equivalently we can regard ω as an $n \times n$ matrix of complex-valued 1-forms ω_{ij} on P satisfying $\omega_{ji} = -\overline{\omega}_{ij}$, and similarly for the associated curvature 2-forms Ω_{ij}.

We will denote by $\mathcal{R}$ the ring of complex-valued polynomial functions on the vector space $L(\mathbf{U}(n))$. Using the usual basis for the $L(\mathbf{U}(n))$, we can identify $\mathcal{R}$ with complex polynomials in the $2n(n-1)$ variables X_{ij}, Y_{ij} $1 \leq i < j \leq n$ and the n variables Y_{ii} $1 \leq i \leq n$. Z will denote the $n \times n$ matrix of elements in $\mathcal{R}$ defined by $Z_{ij} = X_{ij} + \sqrt{-1}Y_{ij}$, $Z_{ji} = -X_{ij} + \sqrt{-1}Y_{ij}$, and $Z_{ii} = \sqrt{-1}Y_{ii}$ for $1 \leq i < j \leq n$. We can also regard Z as representing the general element of $L(\mathbf{U}(n))$, and we will write $Q(Z)$ rather than $Q(X_{ij}, Y_{ij})$ to denote elements of $\mathcal{R}$. The adjoint action of the group $\mathbf{U}(n)$ on its Lie algebra $L(\mathbf{U}(n))$ is now given by $\mathrm{ad}(g)(Z) = gZg^{-1}$, just as in the $\mathbf{SO}(n)$ case above, and as in that case we define the adjoint action of $\mathbf{U}(n)$ on $\mathcal{R}$ by $(\mathrm{ad}(g)Q)(Z) = Q(\mathrm{ad}(g)Z)$. As before we denote by $\mathcal{R}^{\mathrm{ad}}$ the subring of $\mathcal{R}$ consisting of ad invariant polynomials. Once again we can substitute the curvature forms Ω_{ij} for the Z_{ij} in an element $Q(Z)$ in $\mathcal{R}$, and obtain a differential form $Q(\Omega)$ on P; if Q is homogeneous of degree d in its variables then $Q(\Omega)$ is a $2d$-form. The same argument as in the $\mathbf{SO}(n)$ case shows that if $Q \in \mathcal{R}^{\mathrm{ad}}$ then $Q(\Omega)$ is the pull-back of a uniquely determined form $Q(\Psi)$ on M. Using the Bianchi identity, Chern showed that $dQ(\Psi) = 0$, (cf. [MS], p.297) so $Q(\Psi)$ represents an element $[Q(\Psi)]$ in $H^*(M)$, the complex de Rham cohomology ring of M. If we use a different connection ω' on P with curvature matrix Ω' then we get a different closed form $Q(\Psi')$ on M with $\Pi^*(Q(\Psi')) = Q(\Omega')$. What is the relation between $Q(\Psi')$ and $Q(\Psi)$? Weil provided Chern with the necessary lemma: they differ by an exact form, so that $[Q(\Psi)]$ is a well-defined element of $H^*(M)$, independent of the connection. We will denote it by $\hat{Q}(P)$. (Weil's lemma can be derived as a corollary of the fact that $Q(\Psi)$ is closed. For the easy but clever proof see [MS] p.298).

If $h : M' \to M$ is a smooth map, then a connection on P "pulls-back" naturally to one on the $U(n)$-bundle $h^*(P)$ over M'. The curvature forms likewise are pull-backs, from which it is immediate that $Q(h^*(P)) = h^*(Q(P))$. In other words, $Q \mapsto \hat{Q}$ is *a map from* $\mathcal{R}^{\text{ad}}$ *into* $\text{Char}(\mathbf{U}(n))$. It is clearly a ring homomorphism, and in recognition of Weil's lemma Chern called it the Weil homomorphism, but it is more commonly referred to as the Chern-Weil homomorphism.

For $\mathbf{U}(n)$ the ring $\mathcal{R}^{\text{ad}}$ of ad-invariant polynomials on its Lie algebra has an elegant and explicit description that follows easily from the diagonalizability of skew-hermitian operators and the classic classification of symmetric polynomials. Extend the adjoint action of $\mathbf{U}(n)$ to the polynomial ring $\mathcal{R}[t]$ by letting it act trivially on the new indeterminate t. The characteristic polynomial $\det(Z + tI) = \sum_{k=0}^{n} \sigma_k(Z)t^{n-k}$ is clearly ad-invariant, and hence its coefficients $\sigma_k(Z)$ belong to $\mathcal{R}^{\text{ad}}$. Substituting a particular matrix for Z in $\sigma_k(Z)$ gives the kth elementary symmetric function of its eigenvalues; in particular $\sigma_1(Z) = \text{trace}(Z)$ and $\sigma_n(Z) = \det(Z)$. Now if $P(t_1, \dots, t_n) \in \mathbb{C}[t_1, \dots, t_n]$ then of course $P(\sigma_1[Z], \dots, \sigma_n[Z])$ is also in $\mathcal{R}^{\text{ad}}$. In fact, $\mathcal{R}^{\text{ad}} = \mathbb{C}[\sigma_1, \dots, \sigma_n]$, i.e., $P(t_1, \dots, t_n) \mapsto P(\sigma_1[Z], \dots, \sigma_n[Z])$ *is a ring isomorphism*. From this fact, together with Ehresmann's explicit description of the homology of complex Grassmannians, Chern was easily able to verify that the *Chern-Weil homomorphism is in fact an isomorphism of* $\mathcal{R}^{\text{ad}}$ *with* $\text{Char}(\mathbf{U}(n))$. For technical reasons it is convenient to renormalize the polynomials $\sigma_k(Z)$, defining $\gamma_k(Z) = \sigma_k\left(\frac{1}{2\pi i}Z\right)$. Then we get a $\mathbf{U}(n)$-characteristic class $c_k = \hat{\gamma}_k$ of dimension $2k$, called the kth *Chern class*, and these n classes $c_1, \dots, c_n$ are polynomial generators for the characteristic ring $\text{Char}\mathbf{U}(n))$; that is each $\mathbf{U}(n)$-characteristic class c can be written uniquely as a polynomial in the Chern classes.

If $F(Z)$ is a formal power series, $F = \sum_0^\infty F_r$, where F_r is a homogeneous polynomial of degree r, then for finite dimensional spaces, $\hat{F}_r$ will vanish for large r so $\hat{F} = \sum_0^\infty \hat{F}_r$ will be a well-defined characteristic class. Many important classes were defined in this way by Hirzebruch, and Chern used the power series $E(Z) = \text{trace}\left(\exp\left(\frac{1}{2\pi i}Z\right)\right)$ to define the *Chern character*, $\mathbf{ch} = \hat{E}$. It plays a vital rôle in the Atiyah-Singer Index Theorem.

Chern also developed a generalization of the Chern-Weil homomorphism for an arbitrary compact Lie group G. The adjoint action of G on its Lie algebra $L(G)$ induces one on the ring $\mathcal{R}$ of complex-valued polynomial functions on $L(G)$, so we have a subring $\mathcal{R}^{\text{ad}}$ of adjoint invariant polynomials. Substituting curvature forms of G-connections on G-principal bundles into such invariant polynomials Q, we get as above a Chern-Weil homomorphism $Q \mapsto \hat{Q}$ of $\mathcal{R}^{\text{ad}}$ to the characteristic ring $\text{Char}(G)$ (defined with respect to complex de Rham cohomology) and this is again an isomorphism. Of course, for general G the homology of the classifying space $\mathcal{B}_G$ will have torsion, so there will be other characteristic classes beyond those picked up by de Rham theory. Moreover the explicit description of the ring of adjoint invariant polynomials is in general fairly complicated.

Chern left the subject of characteristic classes for nearly twenty years, but then returned to it in 1974 in a now famous joint paper with J. Simons [103]. This paper is a detailed and elegant study of the phenomenon of transgression in principal bundles. Let M be an n-dimensional smooth manifold, $\Pi : P \to M$ a smooth principal G-bundle over M, ω a G-connection in P, and Ω the matrix of curvature 2-forms. Given an adjoint invariant polynomial Q on $L(G)$, homogeneous of degree ℓ, we have a globally defined closed 2ℓ-form $Q(\Psi)$ on M that represents the characteristic class $\hat{Q}(P) \in H^{2\ell}(M)$, and that is characterized by $\Pi^*(Q(\Psi)) = Q(\Omega)$. Chern and Simons first point out the simple reason why $Q(\Omega)$ **must** be an exact form on P. Indeed, by the naturality of characteristic classes under pull-back, $Q(\Omega)$ represents $\hat{Q}(\Pi^*(P))$. But as we saw earlier, $\Pi^*(P)$, the "square" of the bundle P, is a principal G-bundle over P with a global cross-section, hence it is trivial and all of its characteristic classes must vanish. In particular $\hat{Q}(\Pi^*(P)) = 0$, i.e., $Q(\Omega)$ is exact.

They next write down an explicit formula in terms of Q, ω, and Ω for a $2\ell-1$ form $TQ(\omega)$ on P, and show that $dTQ(\omega) = Q(\Omega)$. $TQ(\omega)$ is natural under pull-back of a bundle **and** its connection. Now suppose $2\ell > n$. Then $Q(\Psi) = 0$, so of course $Q(\Omega) = 0$, i.e., in this case $TQ(\omega)$ is closed, and so defines an element $[TQ(\omega)]$ of $H^{2\ell-1}(P)$. If $2\ell > n+1$ Chern and Simons show this cohomology class is independent of the choice of connection ω, and so defines a "secondary characteristic class". However if $2\ell = n+1$ then they show that $[TQ(\omega)]$ does depend on the choice of connection ω.

They now consider the case $G = \mathbf{GL}(n, \mathbb{R})$ and consider the adjoint invariant polynomials Q_k defined by $\det(X + tI) = \sum_{i=0}^{n} Q_i(X)t^{n-i}$. Taking $Q = Q_{2k-1}$ they again show $Q(\Omega) = 0$ provided ω restricts to an $\mathbf{O}(n)$ connection on an $\mathbf{O}(n)$-subbundle of P, so of course in this case too we have a cohomology class $[TQ(\omega)]$. The specialize to the case that P is the bundle of bases for the tangent bundle of M and ω is the Levi-Civita connection of a Riemannian structure. Then $[TQ(\omega)]$ is defined, but depends in general on the choice of Riemannian metric. Now they prove a remarkable and beautiful fact–$[TQ(\omega)]$ is invariant under conformal changes of the Riemannian metric! Such conformal invariants have recently been adopted by physicists in formulating so-called conformal quantum field theories.

Chern also returned to the consideration of characteristic classes and transgression in another joint paper, this one with R. Bott [92]. Here they consider holomorphic bundles over complex analytic manifolds, where there is a refined exterior calculus, using the ∂ and $\overline{\partial}$

operators, and they prove a transgression formula for the top Chern form of a Hermitian structure with respect to the operator $i\partial\bar{\partial}$. This work has applications both to complex geometry (especially the study of the zeros of holomorphic sections), and to algebraic number theory. In recent years it has played an important rôle in papers by J.M. Bismut, H. Gillet, and C. Soulé.

References

[ABP] M.F. Atiyah, R. Bott, and V.K. Patodi, *On the heat equation and the index theorem*, Invent. Math., [19], (1973), 279-330.

[CSP] *Shiing-shen Chern Selected Papers*, Four Volumes, Springer-Verlag, New York, Vol. I, (1978), Vol. II, III, IV (1989).

[G] P.A. Griffiths, *Some Reflections on the Mathematical Contributions of S.S. Chern*, in Volume I of [CSP], xiii-xix.

[MS] J. Milnor and J.D. Stasheff, *Characteristic Classes*, Ann. of Math. Studies 76, Princeton University Press, Princeton, 1974.

[SCW] *Selected Chinese Writings of Shiing-shen Chern* (in Chinese), Science Press, Beijing, (1989).

[W] A. Weil, *S.S. Chern as Geometer and Firend*, in Volume I of [CSP], ix-xii.

S.S. Chern and I

Chen Ning Yang
Institute for Theoretical Physics
State University of New York
Stony Brook, New York 11794-3840

I do not remember whether I had met Prof. S.S. Chern when he was a graduate student of Tsinghua University in Peking (now Beijing) where my father was a mathematics professor, and I was in elementary school. But I do remember how I had met Mrs. Chern for the first time, in early October, 1929, when I was seven years old and she was in junior high school. Her father, Professor Tsen, had been a professor of Mathematics at Tsinghua University already for a number of years, and the Yangs were new comers that fall. The Tsens invited us to their house for dinner and that was when I first made the acquaintence of "big sister Tsen".

Chern was well known as a brilliant graduate student at Tsinghua before he went to Germany for his Ph.D. When he came back to China to teach in 1937, Tsinghua University had, because of Japanese invasion, first moved to Changsha, later to Kunming, combining with the University of Peking and Nankai University to form the wartime National Southwest Associated University. Chern was a brilliant and popular professor, and contributed greatly, together with L.K. Hua, P.L. Hsu and a number of other young professors, to an active atmosphere of mathematical research on the campus. I have very fond memories of my student years on that campus and am deeply grateful for the excellent education I received there (1938-1944)[1].

I probably audited several of Professor Chern's courses in mathematics, but a transcript of my records, which I still have today, shows that I had taken only one course with him, Differential Geometry, in the fall semester of 1940 when I was a junior year student of physics.

I do not remember that course today very distinctly. But one thing does stick to my mind: how to prove that every 2-dimensional surface is conformal to the plane. I knew how to transform the metric into the form $A^2du^2 + B^2dv^2$, but for a long time had not been able to make further progress. When Chern told me to use complex variables and write $Cdz = Adu + iBdv$ it was like a bolt of lightning which I never later forgot.

The Tsen and Yang families were very close, and it was a great joy for my parents to have been among the "introducers" in the marriage of the Cherns in Kunming in 1939.

As is well known, Chern came to the Institute for Advanced Study in Princeton in 1943 and in the next two years transformed global differential geometry with his work on the Chern class. Although throughout the 1950's and 1960's we had seen each other in Chicago, in Princeton, and in Berkeley many, many times, and I had heard of the great importance of the Chern Class, I did not know what it was. Then in the late 1960's I realized that the mathematics of gauge fields is exactly similar to that of Riemannian geometry. So I went to talk to Jim Simons, who was Chairman of the Mathematics Department at Stony Brook. He told me to read Steenrod's *The Topology of Fibre Bundles*. Now I had met Steenrod earlier. I had bought his house on Carter Road in Lawrenceville near Princeton In 1955. I remembered him as a very noncommunicative person. I found his book to be equally noncommunicative, if not more so, and got nothing from it.

It was only in 1975, when Simons gave a series of talks to us at the Institute for Theoretical Physics at Stony Brook that I finally understood the basic ideas of fibre bundles and connections on fibre bundles. Subsequently, T.T. Wu and I re-examined Maxwell's theory and nonAbelian gauge field theory, using fibre bundle viewpoints. Also H.S. Tsao and I, after some struggles, finally understood the very general Chern-Weil theorem.

It is hard to describe the joy I had in understanding this profoundly beautiful theorem. I would say the joy even surpassed what I had experienced upon learning, in the early 1960's, Weyl's powerful method of computing characters for the representations of the classical groups, and upon learning the beautiful Peter-Weyl theorem. Why? Perhaps because the Chern-Weil theorem is more geometrical.

But it was not just joy, there was something more, something deeper: after all, what could be more mysterious, what could be more awe-inspiring, than to find that the basic structure of the physical world is intimately tied to deep mathematical concepts which were developed out of considerations rooted only in logic and in the beauty of form? I had tried to describe my feelings in a talk[2] at the second Marcel Grossman meeting in honor of the hundredth anniversary of the birth of Albert Einstein:

In 1975, impressed with the fact that gauge fields are connections on fibre bundles, I drove to the house of Shiing Shen Chern in El Cerrito, near Berkeley. (I had taken courses with him in the early 1940's when he was a young professor and I an undergraduate student at the National Southwest Associated University in Kunming, China. That was before fiber bundles had become important in differential geometry and before Chern had made history with his contributions to the generalized Gauss-Bonnet theorem and the Chern classes.) We had much to talk about: friends, relatives, China. When our conversation turned to fiber bundles, I told him that I had finally learned from Jim Simons the beauty of fiber-bundle theory and the profound Chern-Weil theorem. I said I found it amazing that gauge fields are exactly connections on fiber boundles, which the mathematicians developed <u>*without reference to the physical world.*</u> *I added "this is both thrilling and puzzling, since you mathematicians dreamed up these concepts out of nowhere." He immediately protested, "No, no. These concepts were not dreamed up. They were natural and real."*

Deep as the relationship is between mathematics and physics, it would be wrong, however, to think that the two disciplines overlap that much. They do not. And they have their separate aims and tastes. They have distinctly different value judgements, and they have different traditions. At the fundamental conceptual level they amazingly share some concepts, but even there, the life force of each discipline runs along it own veins. (Compare Figure).

Around New Year's Day, 1949, the Cherns passed by Chicago on their way from Nanking (now Nanjing) to Princeton. I went to their hotel, the Windemere, to see them. That was when I first met their two children, Paul and May. I remember vividly May, dressed in furry white, crawling all over the carpet of their hotel room. I did not later have the pleasure of serving as an "introducer" in her marriage to Paul C.W. Chu, but I shared the great happiness of the Cherns when the announcement was made in February 1987 from Chu's Houston laboratory that he and his colleagues had found a substance that superconducts at above liquid nitrogen temperatures, a truly world-shaking discovery that ushered in a whole new branch of physics, and promised enormous future industrial applications.

Mathematics Physics

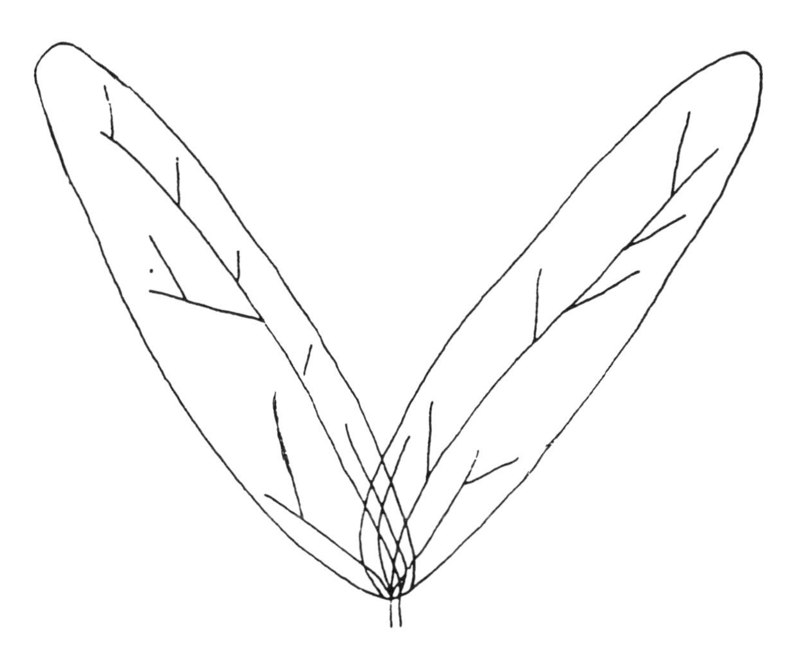

Figure

1. Chen Ning Yang, *Selected Papers 1945-1980 with Commentary*, p.3 (Freeman, 1983).

2. Chen Ning Yang in *Selected Studies; Physics-Astrophysics, Mathematics, History of Science*, ed. T.M. Rassias and G.M. Rassias, P. 139(North Holland, 1982). Also in Physics Today, 33 42(1980) but with the 2-leaves diagram left out.

Ode to Chern Classes

(originally published in <u>The Seventies</u>, Feb., 1983)

天衣豈無縫
匠心剪接成
渾然歸一体
廣邃妙絕倫

造化愛几何
四力纖維能
千古寸心事
歐高黎嘉陳

In a six-year period during the Sino-Japanese War, I had a chance to study with Professor S. S. Chern at the "Southwest Associated University" in Kun Ming since I was pursuing my undergraduate and graduate studies at that time.

Later, Professor Chern went to the United States and did his research in the Institute for Advanced Study at Princeton University. In 1944, he published an internationally acclaimed paper which has opened up new dimensions in the fields of Differential Geometry and Topology.

In this paper, Chern Classes, a very significant concept of the Fibre Bundle Theory, was established. Not only is this concept a remarkable breakthrough in scientific research, it is also a theoretical construct of beauty and wonder: The pieces of a dissected complete manifold can be beautifully and flawlessly stitched together to form a totality in its original state. Learning of this mystery in 1975, I could not help expressing my admiration for his achievements with great awe and wonder.

Why would a physicist like me attempt to study Chern Classes? There is definitely a reason to it. In the past three decades, findings in Physics have discovered that there are four types of force in the natural world: nuclear force, electromagnetic forec, weak interaction, and gravitation. These four forces and their energy are actually Gauge Fields.

In fact, the equations of Gauge Fields are derived and expanded upon from electromagnetic equations laid down in the nineteenth century. It is astonishing to find that these equations are closely related to the Fibre Bundle Theory in Mathematics. Furthermore, the relation between the electromagnetic equations and Chern Classes was revealed in 1974. Due to this important discovery, physicists have learned the importance of Chern Classes. The fact that the explanations of the natural forces have to be found in the Fibre Bundle Theory in Geometry is still a conundrum in the field of scientific research.

Professor Chern's contribution to and accomplishments in the area of Geometry are comparable to those made by Euclid (about 300 B.C.), Gauss (1777-1855), Riemann (1826-1866), and Cartan (1869-1951), who are the geniuses in the history of Mathematics.

THE CHINESE UNIVERSITY OF HONG KONG 香港中文大學

HONG KONG TEL.: 609-6146 FAX: (852) 603-5616 BITNET: cnyang@cucsd.cs.cuhk.hk

C. N. YANG

省身師：

曾国藩謂讀書人成功秘訣是"聰明早達太平壽考"。吾師今達八十歲高齡，四項條件俱備矣。(四十五年無核战亦可謂太平盛世)。謹祝

万事如意

振宁 九〇年七月五日

"聰明早達，太平壽考"

Ingenium et prodigiosum

Longaevitas intra otium

S.S. Chern as Geometer and Friend[1]

André Weil

The friends and colleagues of S. S. Chern who have planned this volume in his honor have asked me for a contribution. Such an invitation is also an honor and could not easily be declined. At the same time, while I have no doubt that future historians of differential geometry will rank Chern as the worthy successor of Elie Cartan in that field, I do not feel competent to give an assessment of his work, nor called upon to do so, since the best part of it, or at least a very representative selection from it, is reproduced in this volume and speaks for itself. All I can do on this occasion is to evoke memories from a friendship of long standing - a friendship which has been among the most valuable ones, personally and scientifically, that I have been privileged to experience.

I must confess that even Chern's name was unknown to me in 1942 when I was asked to review his *Annals of Mathematics* paper on integral geometry ([18] of his bibliography). As I found out later, I had met him briefly in Paris during the year that he spent there in 1936-1937; I was then on the Faculty in Strasbourg and came to Paris regularly to attend the Julia seminar, organized by my friends and myself, which

[1]S. S. Chern, Selected Papers, Vol. I, ©Springer-Verlag, 1978, pp. ix-xiii, by kind permission of Springer-Verlag

met there every other week. The topic of the seminar for that year was Elie Cartan's work; of course it was of special interest to Chern. However, I was invited to spend the second term (January to April 1937) at the Institute for Advanced Study in Princeton, and did not come back until the fall. Thus I did see him in the fall of 1936, but formed no acquaintance with him; to me he was just an anonymous young man from China, soon lost sight of and forgotten.

Five years later Chern was hardly a beginner any more, but somehow none of his published papers had attracted my attention; in part they had appeared in journals which were not even accessible to me at that time. Having left France early in 1941, I was that year at Haverford and had just written, in collaboration with Carl Allendoerfer, a paper on the Gauss-Bonnet formula. My work on the Haar measure and invariant measures in homogeneous spaces, and the interest I was taking in de Rham's work, had brought me close to "integral geometry," which had been a favorite subject for Blaschke and his students in the thirties. This made it natural for *Mathematical Reviews*, then in its infancy, to send me Chern's article [18] for review.

As I duly mentioned, the paper had some weak points. Nevertheless, it lifted the whole subject at one stroke to a higher plane than where Blaschke's school had left it, and I was impressed by the unusual talent and depth of understanding that shone through it. I tried to indicate this in my review, and also pointed it out to Hermann Weyl. As it happened, Veblen was well aware of Chern's work on projective differential geometry, and he and Weyl were considering an invitation to Chern to come to the Institute in Princeton, in spite of the enormous practical problems which this involved; since Pearl Harbor, war was everywhere; a trip from China to America was more than adventurous, it was risky; merely to obtain the necessary visas and priorities on airplanes required setting the whole diplomacy of the USA in motion. Needless to say, none of this fell to my share; I was a helpless refugee myself, officially classified as "enemy alien." All I could do was to express to Hermann Weyl my warm approval of the whole plan, and it is a matter of no little self-satisfaction to me to think that thus, in a small way, I may have contributed to Chern's coming to Princeton in 1943.

When he reached America, I was still not too far from Princeton, and he soon came to visit me. As we found out at once, we had many interests in common. Both of us had been deeply impressed by Elie Cartan's work and by the masterly presentation that Kähler had given of part of that work in his *Einführung in die Theorie der Systeme von Differentialgleichungen*; both of us had known Kähler in Hamburg. We were both interested in the Gauss-Bonnet formula. We were both beginning to realize the major role which fibre-bundles were playing, still mostly behind the scenes, in all kinds of geometrical problems. Better still, we seemed to share a common attitude towards such subjects, or rather towards mathematics in general; we were both striving to strike at the root of each question while freeing our minds from preconceived notions about what others might have regarded as the right or the wrong way of dealing with it.

Chern and I had been particularly intrigued by the little which was then known about characteristic classes (for which no name had been devised yet). Some mystery seemed to hide behind the fact that some Stiefel-Whitney classes were only defined modulo 2. I was able to tell Chern about the "canonical classes" in algebraic geometry, as introduced in the work of Todd and Eger. Their resemblance with the Stiefel-Whitney classes was apparent, while they were free from the defect (if it was one) of being defined modulo 2; their status, however, was somewhat uncertain, since that work had been done in the spirit of Italian geometry and still rested on some unproved assumptions. As to Pontrjagin classes, they had not yet been heard of at that time.

Such were the topics which came up during Chern's first visit and on subsequent occasions, which we sought to renew as often as we could. Very soon, as every geometer knows, they were completely transformed at the hands of Chern, first with his proof of the Gauss-Bonnet formula and then with his fundamental discovery of the role played by complex or quasicomplex structures in global differential geometry. Any comment would be superfluous; I will merely point out what can now be realized in retrospect about Chern's proof for the Gauss-Bonnet formula, as compared with the one Allendoerfer and I had given in 1942, following in the footsteps of H. Weyl and other writers. The latter proof, resting on the consideration of "tubes," did depend (although this was not apparent at the time) on the construction of a sphere-bundle, but of a non-intrinsic one, viz., the transversal bundle for a given immersion in Euclidean space; Chern's proof operated explicitly for the first time with an intrinsic bundle, the bundle of tangent vectors of length 1, thus clarifying the whole subject once and for all.

Chern and I had then to part for a while; I left for Brazil at the end of 1944, while he had to wait until 1946 before being able to go home to his family which he had been compelled to leave behind when first coming to America. There was not much communication between us during those years. My own ideas about fibre-bundles in algebraic geometry were maturing slowly, under the influence of Chern's work on complex manifolds. I knew that he was organizing an Institute of Mathematics in Nanking; I was also watching the political and military developments in China, with increasing anxiety for his fate. In 1947 I came to Chicago, where Marshall Stone had thoroughly reorganized the department of mathematics; he wrote to Chern, offering a visiting appointment. In the Fall of 1948, the civil war was coming closer and closer to Nanking; Veblen and Weyl, obviously feeling as I did about Chern, sent him an invitation to the Institute, backed up, as he later told me, by a friendly cable from Oppenheimer. Chern realized that he had to act quickly; he sent two cables, one to me and one to Princeton, telling us that he was coming to United States.

I have criticized, sometimes severely, the American system of higher education; but I have often quoted the episode of Chern's second coming to America as an example of the flexibility which is perhaps its best feature. When Chern's cable reached me, Stone was travelling in South America. A single exchange of cables with him was enough; on his suggestion and mine, the professors in the department voted to ask for Chern's immediate appointment as a full professor. In the following months, there was some difficulty with the administration; obviously they thought that Chern, as a refugee, could be acquired more cheaply; I knew this attitude, which I had personally experienced during the war. It took Stone's coming back, a threat of resignation from him, and a personal appeal to Robert Hutchins to clinch the matter. Hutchins was in bed with the flu; also, he was technically on leave from the University. But the appointment went through, to take effect in the summer of 1949. In the meanwhile, the Institute in Princeton had provided the funds for him to come over with his family and stay in Princeton until his appointment would start in Chicago.

So it came about, in January 1949, that I could welcome Chern in Chicago at the railway station (not the airport; one still had that option) when he stopped there on his way to Princeton. On that day I met his wife and children for the first time, and remember the occasion vividly. Chern, in his fur cap, looked very much the Manchurian general. But to me the most unforgettable sight was his daughter May, a small girl not yet two years old, all wrapped up in white furs; nothing more lovely could have been imagined.

Thus we became colleagues in Chicago, and remained so for the better part of the decade that followed. We were also close neighbors, housed in the same Faculty building; the University had just built it across the Midway. These were fruitful years scientifically, for him and for me. Fiber-bundles, complex manifolds, homogeneous spaces were prominent among our interests; we discussed them in our offices in Eckhart Hall, or at home, or, better still, on long or short walks through the neighborhood parks, where it was still possible to take a walk and come out alive. Relations with colleagues and graduate students were cordial; visitors, American or foreign, succeeded one another in a steady flow, for longer or shorter stays. With Ed Spanier's appointment a true topologist was added to the team. A quick look through Chern's list of publications in those years, or through mine, will bear witness to the stimulating influence which this scientific atmosphere had on our work.

The time came when circumstances persuaded both Chern and myself to seek elsewhere, among other things, a better climate and more pleasant physical surroundings. As we had sometimes jokingly predicted, he found them by moving closer to China, and I to France. This did not slacken the bonds friendship, but it is only natural that from then on we followed each other's work less closely, even though we did arrange to get together at not too distant intervals. It is entirely to him and to the ties he hand kept up with his colleagues in China that I owed my invitation there in the fall of 1976-an unusual experience which left on me a deep impression. But, rather than commenting upon such personal matters, or upon Chern's work of the last fifteen years (which others would be more competent to discuss, and whose value is recognized by all), it is perhaps appropriate to conclude with a few words about the place of geometry in mathematics - the mathematics of today and presumably also the mathematics of tomorrow.

Obviously everything in differential geometry can be translated into the language of analysis, just as everything in algebraic geometry can be expressed in the language of algebra. Sometimes mathematicians, following their personal inclination or perhaps misled by a false sense of rigor, have turned their mind wholly to the translaton and lost sight of the original text. It cannot be denied that this has led occasionally to work of great value; nevertheless, further progress has invariably involved going back to geometric concepts. The same has happened in our times with topology. Whether one considers analytic geometry at the hands of Lagrange, tensor calculus at those of Ricci, or more modern examples, it is always clear that a purely formal treatment of geometric topics would invariably have killed the subject if it had not been rescued by true geometers, Monge in one instance, Levi-Civita and above all Elie Cartan in another.

The psychological aspects of true geometric intuition will perhaps never be cleared up. At one time it implied primarily the power of visualization in three-dimensional space. Now that higher-dimensional spaces have mostly driven out the more elementary problems, visualization can at best be partial or symbolic. Some defree of tactile imagination seems also to be involved. Whatever the truth of the matter, mathematics in our century would not have made such impressive progress without the geometric sense of Elie Cartan, Heinz Hopf, Chern, and a very few more. It seems safe to predict that such men will always be needed if mathematics is to go on as before.

Shiing-Shen Chern as Friend and Mathematician, A Reminiscence on the Occasion of His 80th Birthday

Wei-Liang Chow

Department of Mathematics

Johns Hopkins University

I met Shiing-Shen Chern for the first time in Hamburg sometime in the fall of 1934. We were at that time both students at the Hamburg University; Chern was at that time studying with Blaschke, while I was attending the lectures of Artin. I purposely use the rather vague expression "attending the lectures of Artin" in order to indicate the tentative nature of my sojourn in Hamburg at that time; more correctly, I should say that I was at that time strictly speaking a student at the University of Leipzig, but I decided to stay in Hamburg for personal reasons, as I shall explain later, and I took the opportunity to learn something by attending the lectures of Artin. In order to explain this rather strange lack of definiteness in my mathematical aspirations and also in order to understand the importance of Chern's influence at a critical juncture later in my life, let me say a few words about my rather unusual educational background.

Except for a very brief period, I never attended schools or colleges in China. Beginning at the age of five (1916) I was taught the standard Chinese classics by an old Chinese tutor and at the age of eleven I was taught to read and write English. However, I discovered very soon that the ability to read English provided me with the opportunity to acquire the knowledge about almost any subject I wanted to learn. Since the curricula in most Chinese universities at that time were modeled after those in the American universities and many of them often used books written by American professors, it was not difficult for me to find out the most commonly used text books in America on most subjects. Thus in this way I taught myself all sort of subjects from mathematics and physics to history and economics. This situation lasted from 1924 to 1926 when I succeeded in persuading my father to send me to study in the United States. At that time my main interest was political economy, and economics was still my major subject of study when I entered the University of Chicago in October 1929. However, during the next two years I began to have some serious doubts about taking economics as my major.

Already in my childhood I always wanted to be an electric engineer, although I did not really know at that time what that implies. Now, as my doubts about the wisdom of majoring in economics increased, I decided to change my major to physics, hoping that it would eventually lead me to engineering. Thus, when I graduated at the University of Chicago in 1930, my major was in physics. At about the same time I happened to read the book called *Pure Mathematics* by the famous English mathematician Hardy. This book opened the door to mathematics for me, although I was at that time still studying applied mathematics, hoping eventually to study physics. In summer 1931 I discussed studying mathematics with a graduate Chinese mathematics student who got his Ph.D. at Chicago and then spent a year in Princeton. He was very enthusiastic about Princeton (he attended the lectures of John Von Neumann there) and he advised me to go to

Princeton or even better to go to Goettingen in Germany which he thought was then the world center for mathematics. Therefore, with only a vague idea of studying mathematics, I went to Goettingen in October 1932. Although I had previously taken a course in German at the University of Chicago, it took me about three months to learn the German language sufficiently to enable me to understand the lectures. However, by that time, at the beginning of 1933, something happened in the German politics which would soon change drastically not only the university at Goettingen, but the entire Germany and in fact eventually the entire world, namely Hitler and his Nazi party came to power. Not knowing anything about the German politics, I was surprised to hear some students murmuring that Hermann Weyl would very probably leave Goettingen, which Weyl did that summer. At about the same time first Richard Courant and then Emmy Noether also left Goettingen. Thus the world mathematics center I hoped to come to study was essentially depleted.

Thus in the summer of 1933 I had to find a German university to replace Goettingen. The summer before in Chicago I had taken a course in modern algebra, in which the then relatively new text book called *Moderne Algebra* by van der Waerden was used, and I was very much impressed by the elegance of the presentation of the subject in that book, and upon learning that van der Waerden was a professor at the university in Leipzig, I went to Leipzig, hoping to study with van der Waerden. It was a stroke of luck for me that I went to Leipzig at just about the time when van der Waerden was writing the beginning of his series of papers entitled *Zur algebraischen Geometrie*, and as a result I was introduced to a subject I never heard of before. Van der Waerden was very kind to me; he told me to study the books of Severi and also the older books of Bertini and Enriques. Also, van der Waerden has the unusual talent of explaining even the most complicated mathematical theory in rather simple terms and he made me feel that my lack of knowledge of some important subjects in mathematics was not important as long as I was willing to learn; in fact, for the first time in my life I began to feel that I made the right choice to study mathematics.

I went to Hamburg in the summer of 1934 for summer vacation and there I met a young lady, Margot Victor, who eventually became my wife. It was a case of love at first sight, and I actually proposed marriage to Margot within a week after we met. Realizing that my feeling of love could not be expected to be reciprocated within a short period of time, I decided to stay in Hamburg to woo Margot. At the same time I would take this opportunity to attend the lectures of Artin, hoping to learn something about algebraic number theory, while at the same time I continued my study of algebraic geometry. (The German university system was very flexible at that time and allowed me the freedom of "wandering" from one university to another.) In time Chern and I became good friends, although we did not have much mathematical contacts with each other at that time. At the beginning of 1936, I went back to Leipzig to finish my dissertation with van der Waerden. When on July 10th, 1936 I married Margot Victor in Hamburg, Chern was present at the reception given by Margot's parents. (See the picture taken at that reception.)

The next time I saw Chern was some time in 1939 or perhaps in 1940 (I do not remember exactly), when he came to Shanghai for a visit. I learned from him that he could not even land in Shanghai when he came back to China in 1937 because of the fighting around Shanghai. I did not know what was the reason for his visit to Shanghai, but we did discuss the possibility of issuing a volume of the journal of the Chinese Mathematical Society. By that time I was forced by circumstances to abandon my mathematical research; after all I had to support not only my wife and two children, but also my parents-in-law, who were forced almost penniless to leave Germany by the racial policy of the

Nazi regime. Although my father was normally a very well-to-do man (after all he supported my study in the United State and Germany for almost nine years), his business and investments suffered serious set backs from the war and I realized that I had to stand on my own and earn a living by doing whatever business I could find.

The next time I met Chern was in the spring of 1946. The war was over by that time, and Chern had just come back from a successful journey of two years at the Institute for Advanced Study in Princeton, and he was called upon to organize a new institute of mathematics for the Academia Sinica in Nanking. On my part I had been practically entirely out of any mathematical activities for almost a decade; the last paper I wrote (1938) was inspired by reading an old paper on thermodynamics by Caratheodory, and although van der Waerden accepted the paper with some words of praise, including some kind comments by Caratheodory, but it was so long ago that I almost forgot about it. The contrast between us in our mathematical positions was very clear. Chern was by that time an established mathematician of international reputation, while I was at best a post-doctoral student, with the additional burden of almost a decade of neglect. My plan for the future, in so far one could make future plans in those confusing and uncertain days in China, was to develop an import and export business through my business connection with an uncle of my wife, Hans Victor, who was a successful business man in Germany before the Nazi regime came to power and was forced almost penniless to leave Germany at the age of sixty, but who nevertheless later built up his business in America. Realizing that I could not rely on what would normally be my inheritance from my father and observing also the uncertainty of any normal academic condition in China, I had to find a way to support not only my wife and two children, but also my mother-in-law (my father-in-law had died in the mean time).

Chern realized that in the rather confused condition in China after the war what was mostly needed by mathematicians in China was recent literature in their own fields of research. For this reason Chern collected a large number of reprints from mathematicians not only in his own fields of research, but also in related fields. For my part I was very happy to find among his collection of reprints most of the recent papers of Oscar Zariski who apparently had carried the modern algebraic geometry far beyond the stage reached by van der Waerden before the war. Chern also told me about the important work of Andre Weil, whose work would probably be published in book form in the near future. It was clear to me at that time that if I ever wanted to get back to mathematics, my first task was to study the papers of Zariski, which Chern very kindly lent to me. Chern also suggested that I spend a year at the Institute for Advanced Study in Princeton in order to get myself into the stream of modern mathematics, and he expressed the opinion that in spite of the loss of my past ten years, it was not too late for me to get started again in mathematics. In fact Chern even wrote a letter to Lefshetz, suggesting that an invitation be extended to me to visit Princeton.

It was clear to me that I had to make a fundamental decision, probably the most basic one in my life, whether I should quit my business activities and try to get back into mathematics. After all, I was then thirty five years old and it was more than ten years after I got my doctoral degree and I had hardly done anything since then. It seemed that I probably had missed my chance to be a mathematician, and I was so discouraged by this situation that I was ready to give up mathematics. However, the conversation with Chern had changed my mind and renewed my hope to stay in mathematics. I discussed the problem with Margot, who showed great understanding for my wish to go back to mathematics; she felt that I should follow my own inclination and judgement, and that whatever I decided to do, she would adjust her activities accordingly. After all, she had worked in her uncle's office for several years before her marriage and she could go back to work again if it would become necessary. We had at that time saved enough money from my past business activities to pay for our passages to the United States and to live for a year or two without any additional income, if necessary. We realized that we were taking a great risk, since the possibility of my achieving even a moderate degree of success in this new endeavor at that stage in my life was by no means assured, but we also felt that sometimes in life one must take bold actions. (As to my mother-in-law, I would provide her with enough funds so that she could go to Capetown in South Africa, where a sister of Margot lived, so that she would not be entirely alone.) Once we had made this decision, the next step was for me to wind up my business and put all my assets in liquid form; and while I waited to get the necessary passports and visa in order, I devoted my entire time and effort to mathematics, in particular to study the papers of Zariski. Originally we intended to leave for the United States sometime in the fall of 1946, but all the arrangements I had to make (plus the fact that I had to attend to some of business matters of my father who was old and in very poor health) delayed our departure until March of 1947.

I arrived in Princeton at the beginning of April in 1947 and was promptly admitted as a temporary member in the Institute for Advanced Study, thanks to the letter of Chern to Lefschetz. It was too late for me to get a stipend even for the forthcoming year, which was not too important for me at that time, but we (i.e. I and my family) could live in the Institute housing, which was important, for there was then a great shortage of housing due to the war. In the spring of 1948 I learned that van der Waerden was visiting at the Johns Hopkins University and I went there to see him. It turned out that there was a vacancy at Hopkins, and van der Waerden, always very kind to me, told me that he would recommend me for this position if I would be interested in it. As a result I went to Hopkins in 1948, and I stayed there until my retirement in 1977. I met Chern again in 1949 at the Institute for Advanced Study, where he was visiting, before he went to Chicago. What happened after that is now public mathematics history; as we all know, Chern was a professor at the University of Chicago until 1960 when he went to Berkeley and he has remained in Berkeley ever since. After Chern had gone to Berkeley, my contact with him was less frequent than when he was in Chicago, owing to the great distance between the east and west coasts, but I did make two summer trips to the Bay area to visit him. Chern has received some of the highest honors which a mathematician can possibly have, and I shall leave this aspect of his career to some of his collaborators or his former students, who are probably in a better position to give an adequate account of the citations leading to these honors. Chronologically, Chern has long passed the normal retirement age (we were both born in 1911). But a mathematician like Chern in a sense never retires. Even after passing his formal retirement age, Chern has been for several years the director of a research institute set up by the National Science Foundation in Berkeley. Also Chern went back to China several times, and I understand there is a research institute at his Alma mater, the Nankai University, of which Chern is the director. I think at the rate Chern is working to promote mathematics in China, he will be not only a leading mathematician of our generation, but also the father of modern mathematics in China.

I think I can best conclude this article by quoting a part of my letter to Chern on the occasion his formal retirement in 1979. Apart from the publication of his Selected Papers, there was an International Symposium in Global Analysis and Global Geometry in Berkeley in his honor, organized by the mathematics department at Berkeley and sponsored by the National Science Foundation. I was invited, and I would certainly have attended this Symposium, if I were not in Germany at that time. Our European trip, our first such trip since we left

Germany more than forty years ago, was planned a long time ago, and our relatives and friends all made plans to fit our time table that it was not possible for us to change our plans at the last minute. Since I could not personally congratulate him on this occasion, I wrote Chern a letter instead. The letter was written on June 20th, 1979, at a place called Schloss Elmau, a beautiful place in the Bavarian Alps, not far from the Austrian border, and I ended my letter with the following statements: "It is befitting that the National Science Foundation and your colleagues at Berkeley have chosen to honor you in this way, as you are no doubt one of the leading mathematicians of our generation. On a more personal level, I shall always remember that it was mainly due to your advice that I returned to mathematics after the war. Without this encouragement on your part at that critical moment in my life, I would probably never have made whatever modest contributions I may have made to mathematics. For this I owe you an enduring personal debt of gratitude."

For Chern Volume

I.M. Singer
Department of Mathematics
Massachusetts Institute of Technology

Last term I gave some extra evening lectures to my class in Geometry and Quantum Field Theory. The younger graduate students needed the geometric background for gauge theories. I happily talked about connections on fibre bundles, curvature, characteristic and secondary characteristic classes. I reviewed the many excellent treatments but finally followed Chern's classic "Geometry of Characteristic Classes." I urged the class to study this short, concrete, elegant and deep paper.

Rereading it brought back vivid memories of Chern's course at The University of Chicago in 1949–1950. Differential forms were not well known then. In fact, a year earlier, Andre Weil had introduced us to differential forms in a seminar (for graduate students AND faculty) in which he recast the calculus of several real variables in the language of forms. Barely familiar with Grassmann algebras, students in Chern's course were amazed and intrigued at Chern's effective use of forms in local and global geometry.

Because I was writing a dissertation at the time, I was a passive participant in the many student discussions trying to understand and absorb the material in Chern's lectures. However, a year later at M.I.T., Warren Ambrose was eager to learn differential geometry. He organized a seminar in which I lectured on my notes of Chern's course. Like many others, I struggled with Chern's "Let p be a point and dp its differential" and finally understood how convenient a notation it was for the identity map and its differential.

My recent evening lectures reminded me of our seminar over forty years ago held in the same room – with blackboards on three walls. We filled all the backboards with the definition of a connection. We hadn't yet learned to first expose the properties of principle bundles and of Lie groups. Chern does that very neatly in the paper cited above. Only long experience has taught me how much goes into making deep ideas simple, as Chern does. To quote him, "This train of ideas is so simple and natural that its importance can hardly be exaggerated."

Under Chern's influence, both Ambrose and I taught courses in differential geometry at M.I.T. Others did the same elsewhere. Books were written and the subject flourished. I need hardly dwell on what we all know: Chern introduced global differential geometry to American mathematics, as well as the use of differential forms in a host of subjects. Geometry is harder to define than most mathematical disciplines. Whatever it is, Chern has shown us for half a century how to do differential geometry. Each time we meet, he shows me something new – and I wish I had been a more attentive student. Why didn't I listen more carefully to his discussion of Backlund transformations, or his exposition of Finsler geometry?

In time, Chern's course and Ambrose's seminar had their effect. I switched from functional analysis to global geometry. My present interest is in Quantum Field Theory. Because the ones of interest in physics are geometric, I continue to be immersed in global geometry. In fact what's missing in physics is some quantized version of geometry that we only dimly perceive in string theory.

I have emphasized Chern's scientific influence. His personal influence has been just as strong. He is very encouraging and gentle with students and younger colleagues. And he runs conferences or institutes with the same directness and elegant simplicity as in his papers.

It's been a privilege to know Chern and his wife, Shih Ning, as friends. In fact it's been fun – the number of Chinese restaurants, modest or fancy, in which we have discussed mathematics, politics, family, and life.

In my view, Chern is one of the great figures in 20th Century mathematics.

Professor S. S. Chern, My Father-in-Law

Paul C. W. Chu

Department of Physics and
Texas Center for Superconductivity
University of Houston
Houston, Texas 77204-5932

Thank you. I would like to thank Professor Yau for giving me this opportunity on such a happy occasion to share with you some of my personal experiences with Professsor S. S. Chern.

S. S. Chern was a household name in Taiwan where I grew up. He had a tremendous influence on those of us in Taiwan who were interested in science and engineering. For me to know *of* him, therefore, was not an accident at all. However, for me to become a member of his family was *quite* an accident. Just think of the small probability that his daughter, May, and I would *ever* meet; one grew up in a well-known family in the United States and the other grew up in a small village in Taiwan, tens of thousands of miles apart from each other.

I vividly remember the first time I met Professor Chern. I was presenting three short papers at the March Meeting of the American Physical Society held in Berkeley in 1968. Prior to my departure for Berkeley, I had a lot of free advice from colleagues at the University of California at San Diego on how to win Professor Chern's blessing for my marriage with his daughter. One of my advisors, for instance, suggested I thoroughly review calculus and differential equations just in case Professor Chern decided to give me an oral exam as a sort of minimum requirement for his future son-in-law.

The occasion for us to meet finally came when May asked me to join her when she picked up her father, who was returning from a lecture in Boston, at the San Francisco Airport. When he arrived, I saw a big, friendly smile on his face, under a broad forehead on an unproportionally large head. He was waving a box of live lobsters from Boston as he rode toward us on the "moving walkway." There went my anxiety. Later, he sent the lobsters to King Tsin, a Chinese restaurant in Berkeley, and asked Chef Liu to cook them for us. They were, indeed, delicious. Since Liu did not want to accept any money for doing the cooking, Chern asked me if he could write a poem for me to execute in Chinese calligraphy to offer to Chef Liu as a token of his appreciation. To hide my weakness in calligraphy, I turned him down.

I guess one just cannot hide one's awkwardness all the time. The most embarrassing moment during that trip came on the final day of my stay at Berkeley when I was invited to attend a Chern family party that evening. I enjoyed it very much. Only much later did I find out from May what happened that evening. Apparently, when Mrs. Chern and May removed the salad plates and forks after the first course of the dinner, they discovered that someone had used the wrong fork. To May's embarrassment, that "someone" was me. To tell the truth, western table etiquette was nowhere to be found in my dictionary in Taiwan. What impressed me the most was that, in spite of my peasant background, the Chern's did not object to their daughter's friendship with me. I was amazed that they were willing to entrust the life of their daughter, their only daughter, to a lowly peasant stranger like me. Perhaps, this was a reflection of their trust in their daughter.

Years later, I discovered another, yet to be authenticated, part of the puzzle. I was told that Professor Chern had asked Professor C. N. Yang, the person you just saw in the slide show, to check me out through my thesis advisor, the late Professor Bernd Matthias. Matthias' answer was characteristically simple: "Paul is bright; but May is brighter." To Professor Chern, this must have been a passing grade, since a physicist may belong to an intelligence ladder second to that of a mathematician, which is transferable to the mathematician's daughter.

Besides lobster, Chern eventually introduced me to many nice but expensive tastes. However, he also demonstrated to me, often in a very subtle way, how to earn a sufficiently respectable living to support such tastes. As young persons usually do, I used to asked him, "How have you managed to become so successful?" In a friendly but indirect way, he basically told me that imitation is not a viable way to success. Instead, one should be himself, not presumptuous, and self-disciplined at all times. One should recognize his own strengths and weaknesses, should follow his own instincts and interests, and should not do something simply for the sake of being fashionable. As a result, whenever one finds something interesting and new, one should stick to it, but with an open mind. He has never been obsessed with money or fame. He always says, "Only ask how much effort you have put into your work; never ask how handsome the reward will be." He has exemplified all of these concepts throughout his life, starting from his days in Germany when he chose the then not-so-fashionable subject of differential geometry as his main field of study.

For more than twenty years, I have taken what May and I call "the annual rejuvenation trip" back to the Chern's home in El Cerrito. This annual trip was especially important prior to the discovery of high temperature superconductivity in 1987. It was a much needed therapy for me, since the road to high temperature superconductivity was extremely lonesome and hazardous, career-wise, before 1987. The Cherns and May provided much of the vital support I needed. After 1987, both

Professor and Mrs. Chern became the most faithful cheerleaders for my research. For quite a long time, the opening question in our phone conversations was, " What is the temperature now?" In fact, I still remember a phone call after 11:00 pm one evening in the summer of 1988, when the Cherns called me from Kennedy Airport in New York City on their way to Europe just to find out our latest results.

Quite aside from the fact that Chern gave me his daughter to be my wife, he has affected my life rather significantly, for better or for worse, either through direct interaction with him or through osmosis. If I have done anything worthwhile in my life, a substantial part of the credit should go to Professor and Mrs. Chern and, of course, to May. I wish Professor and Mrs. Chern many more healthy, happy, and fulfilling years ahead. And finally, it would give me great personal and professional satisfaction if the Chern-Simons invariant were to eventually provide the proper glue to unite his field of differential geometry and my field of high temperature superconductivity.

Thank you.

Shiing-Shen Chern with Admiration as He Approaches His 80th Birthday

Irving Kaplansky
Mathematical Sciences Research Institute
Berkeley

I expect to be best remembered as the person who succeeded Shiing-Shen Chern as Director of MSRI. This happened in the fall of 1984. As I write this, nearly seven years have gone by. At MSRI we are proud and happy that the Founding Director and Director Emeritus has chosen to maintain a strong presence at MSRI, even while he directs another institute: the Nankai Institute in Tianjin, PRC.

All the work stations at MSRI have been named for mathematicians, and one of them is named for Chern. It was the first of the RT's that were acquired as the result of a generous grant from IBM. There was a ceremony naming the work station, and a photograph of the RT appears in the MSRI column in the April, 1989 issue of the Notices of the American Mathematical Society. (Chern politely declined to be photographed with his namesake, so Deputy Director Emery Thomas and I posed instead.)

He was the chairman of the program committee for the 1984–85 MSRI program in Differential Geometry (the other committee members were Blaine Lawson and Is Singer). He was the driving force behind the May 16–20, 1988 workshop on Harmonic Maps and Minimal Surfaces; Bob Osserman (now MSRI's Deputy Director) chaired the workshop and the committee was completed by Richard Schoen. These are the formal aspects of his continued activity at MSRI. Not formally recorded are the inspiration and guidance he generously gives to members, especially young ones.

This volume contains a biography of Chern by Richard Palais and Chun-Lian Terng. Chern himself wrote an autobiographical sketch that appears in the first volume of his Selected Papers. In April, 1991 I had a wide ranging conversation with him as a result of which I can add several details that do not appear in these sources.

My first memory of Chern is a colloquium talk that he gave at the University of Chicago in January, 1946. He was on his way back to China after nearly three years at the Institute for Advanced Study. He was with Bohnenblust at Indiana University over the Christmas holidays; the late Adrian Albert located him at the Indiana Union and invited him to give a talk at Chicago. I remember keenly the energy and mathematical drive in the talk. When I told him my recollection that the topic was his generalized Gauss-Bonnet theorem, he agreed that this was probably right.

From Chicago he took a train to Los Angeles. He spoke at Caltech, where he met E.T. Bell and A.D. Michal. He then went on to speak at Berkeley, meeting G.C. Evans and Hans Lewy.

There followed his return to China and reunion with his family, his second stay at the Institute for Advanced Study in 1949, and his acceptance of a position at Chicago in 1949, where he stayed till 1959. For those ten years we were colleagues and he was one of the people who made that decade such a wonderful "Stone age".

Our work did not overlap very much. By osmosis, however, I caught something of the spirit of modern differential geometry. He was no narrow specialist. I recall one sample of his friendly interest in the work of others when he commented on my paper "Any orthocomplemented complete modular lattice is a continuous geometry".

His decision to leave Chicago for Berkeley in 1959 was a severe blow to Chicago's Mathematics Department, and I joined with everyone in Chicago in deeply regretting it. But in retrospect I see some pluses. Certainly Berkeley got a big boost. And – if he had not moved he would not have been on the scene to prepare (with Cal Moore and Is Singer) the proposal that successfully launched MSRI. In his two years at the helm a flourishing institute was created that I was lucky indeed to inherit.

Here is a vignette that I find noteworthy. His arrival in Berkeley in 1960 (after a leave of absence in 1959–60) marked the beginning of a weekly differential geometry seminar which has continued without interruption to this day. In every semester (or quarter when the quarter system was in operation) he himself has spoken.

In concluding our conversation I asked whether his choice of geometry reflected a natural geometric way of thinking. Not at all, he said; it was because (as he records) his early teachers Chiang and Sun were geometers. In fact, in 1936 Blaschke suggested that he stay on in Hamburg to work on algebraic number theory with Artin and Hecke. He was tempted. Had he done so, the history of mathematics in the twentieth century would have been significantly changed. But instead he went to Paris where he was decisively influenced by Elie Cartan. Differential geometry benefited, but what a loss algebra and number theory suffered!

Some Personal Remarks About S.S. Chern

by Louis Nirenberg
Courant Institute
New York

Chern represents for me, Mr. Differential Geometry. I have considered him the master and chief authority in the field for many years – probably ever since we met. I can't remember exactly when that was, it must have been in the mid-fifties, though it seems to me that I've known him forever. His knowledge of all aspects of Differential Geometry is remarkable: from classical things, including very special questions for two dimensional problems, up to the most general local and global problems; also the interaction of geometry with topology, complex analysis etc.; he is a master of the work and techniques of Elie Cartan. In addition, his research work has covered every aspect of Differential Geometry, and his bag of tools seem to include most fields of mathematics: analytic techniques, algebraic, topological, Lie groups, complex analysis, etc. I will not go into his mathematical work since others, more competent than I will surely do so. In addition, there are excellent short articles by A. Weil(a personal one), P.A. Griffiths, and Chern himself, at the beginning of the volume of his selected papers published by Springer Verlag in 1978. I will just make a few personal remarks.

I have never really spent much time at any university where Chern was – in January 1955 I visited the University of Chicago, but he was away. I have spent several summers in Berkeley, however we didn't interact so much mathematically. Nevertheless, we became friends, also with his wife Shih Ning, and my wife. Chern's unassuming and warm personality quickly makes people feel at ease, at home.

Also professionally, Chern has a special way with people which inspires one to work on problems he suggests. It is a combination of enthusiasm, kindness, and encouragement that one can do something with the problem. It must have been a wonderful experience to have been his student.

Around the time that we met, Chern mentioned to me the problem of proving the integrability of almost complex structure. A. Weil had, earlier, called my attention to the problem and at that time I tried to solve it but made no progress. After Chern brought it up, A. Newlander and I tackled it again and we found a solution.

One of Chern's striking talents is the ability, and liking, for very complicated computations. He is not at all phased by them no matter how long or difficult. I experienced this personally in a small way at the time of our (single) collaboration during the summer of 1969 when I was visiting Stanford. Chern described a problem he was working on with H. Levine (who was not in California that summer). It involved introducing intrinsic pseudo-norms in the real cohomology vector space of a complex manifold. The definition made use of plurisubharmonic functions. They needed some inequality and thought I might be of some help. Chern and I met several times during that summer to discuss the problem. I was able to contribute to the work, but at some point I needed a somewhat complicated inequality in multilinear algebra and I didn't see how to do it. When I asked Chern about this point he said, "Oh yes, it's just – – –" and did it on the spot. The joint paper is

#90 in his bibliography.

A year or so later at a conference at the University of Chicago in which Chern and I participated, I gave a talk on the joint paper. Toward the end, Felix Browder, if memory serves, asked how we came up with the definition of the seminorm. I couldn't resist telling the old story of the travelling scholar during the last century, in Russia, who went from town to town, by coach, giving lectures. As he was approaching a town, he said to his coachman. "Look I've never been to this town, nobody there knows me. I'm tired of giving this lecture, and you've heard it many times. What do you say we change places, and clothes, and you give the lecture?" The coachman agreed, gave the lecture while the scholar relaxed at the back of the hall, and at the end took questions. He was able to answer them all until someone asked one he couldn't handle. At which point he said, "That question is so simple my coachman, will answer it for you", pointing to the back of the hall. As I said this I pointed to Chern. Unlike the travelling scholar, Chern had not been forewarned. He was suddenly startled to find people laughing and all eyes fixed on him. When Browder's question was repeated, his answer was, "Well one fools around". His mathematical "fooling around" has led to many beautiful discoveries.

It was an enormous pleasure for me to work with Chern and I regret that we never collaborated again.

About his love of computation, someone once reported to me that when working on a problem involving Riemannian geometry, if some function (or form) comes up, Chern's automatic reaction is to apply the Laplace operator to it – just to see what properties it might have. In many of his talks I have marvelled at his use of the Cartan formalism to do what seemed to me to be extremely complicated computations.

Over the years, meeting in New York or Berkeley or in China, we have maintained a very warm friendship. It is to Chern that I owe my first contact with mathematicians in China. Also, he and Shih Ning have been wonderful hosts when I visited the Nankai Institute.

In conclusion, I would like to add something known to many: what a special experience it is to go to a Chinese restaurant with Chern (as we always do when we meet). On every such occasion one eats infinitely better than if he had not been there – even if there were other Chinese in the party. I once asked him to tell me his secret. He said, "Well, I just call in advance and say ... " But I can't reveal it. Anyway it wouldn't work for anyone else. Incidently, Shih Ning is a wonderful cook.

Through his research, teaching, and also his personality, Chern has greatly influenced the work of many mathematicians during the last half century. I hope his inspiration will continue for years to come.

S.S. Chern

Felix E. Browder

I first met S.S. Chern in the early 1950's, and certainly by the time I visited the University of Chicago in November 1953 at the invitation of Marshall Stone. Chern was one of the group of senior mathematicians that Stone brought to Chicago in 1947 to transform the Chicago Mathematics Department. (The others of course were Weil, MacLane and Zygmund, and it is curious that by Stone's recollections, the ones he had difficulties appointing were Chern and Weil.) Chern had already become the recognized world leader in global differential geometry, whose fundamental links with the theory of fiber bundles and algebraic topology were set forth in his lecture at the International Congress in Cambridge in 1950. He had played a major role in making exterior differential forms the central tool of this development.

Like many others, I found Chern encouraging and helpful even to those whose mathematical interests had little intersection with his own. My own central concern at the time was the development of the theory of linear elliptic partial differential operators in the context of functional analysis, which had little relation to his concerns. Despite this he was friendly, warm, and courteous, and so he remained for the decades since. He knew the intrinsic value of his own mathematical interests and did not need to press that value ferociously upon others. When new methodologies arose which differed from his own, as in the massive influx of nonlinear p.d.e. methods and a priori estimates as in the work of Calabi, Nirenberg, and especially of Yau and his collaborators, he welcomed the new tendency alongside the more classical approach.

Chern's natural dignity and objectivity caused him to have a strong influence on his colleagues, first at Chicago and later at Berkeley. The school of differential geometry which he created at Berkeley is obvious testimony to this fact. For many years, he was one of the major balance wheels of the Berkeley department, which often seemed to need balancing.

So rarely in my experience did Chern express anger or irritation that when he did, the event was notable. I recall one of little global importance: his anger at the idea of a paper based on the I. Ching which someone wanted to print in the Proceedings of the A.M.S. Global Analysis Summer Institute of 1988. As a self-respecting member of the Chinese intellectual class, he found this a specimen of degradation.

He remains firmly attached to his Chinese roots despite all his early vicissitudes there and his long sojourn in America. He is regarded by all who know him as he wishes to be regarded, as a man of dignity and courtesy, the ideal combination of the mathematician and the Chinese intellectual gentleman. His work has won a permanent place at the center of contemporary mathematics; he is regarded with deep affection and respect by his numerous friends.

Personal Recollection of Chern at Chicago

R. Lashof

Berkeley

My collaboration with Chern began a year or so after my arrival (1954) at Chicago as a young instructor. My thesis was in topology and topological groups. I knew almost nothing about differential geometry but I thought I should learn more about it and asked Prof. Chern for a good book on the subject.

At that time mathematics at Chicago was very abstract. Andre Weil was a potent force, and even if he and MacLane did not get along they both pushed strict algebraic approaches to geometry and topology. I remember Chern giving a lecture in Weil's seminar on "mathematics" in which Chern included an explicit calculation using local coordinates. This was too much for Weil and he blurted out "Chern haven't you learned anything" – meaning of course, hadn't Chern learned to do mathematics a la Bourbaki!

My guess is that Chern was happy to have a young mathematician trained by the likes of Chevalley, Harish-Chandra and Eilenberg show an interest in some of the concrete problems of differential geometry. In any case, he not only suggested several references but also an explicit problem concerning Gaussian curvature and global geometry. Since I knew so little classical geometry, I fell back on the naive approach I had learned from physicists – the so called vector analysis. Chern was able to make sense of my primitive approach and this led to a successful collaboration. It was also a wonderful way to learn differential geometry.

In those days we didn't have computers, and mathematical typists were non-existent. In short, mathematicians typed their own papers. As the junior partner, I offered to type up our first paper, but I was very slow. Chern saw I was struggling, and took the paper home and typed it in one evening.

Chern was considerate with respect to other concerns of young mathematicians. There was a differential geometry conference in Zurich in the summer of 1960. He not only introduced me to the many famous mathematicians attending the conference but also to the local cuisine, the best dishes and the best wine.

I was disappointed when Chern left Chicago for Berkeley. Spanier had already gone to Berkeley taking the young instructor Smale and Ph. D. student Hirsch with him. This group formed the core of the powerful Berkeley faculty in geometry and topology.

For the Chern Volume

Raoul Bott
Department of Mathematics
Harvard University

It must have been during the Spring term of 1949 that I first met Chern. Although the date is now uncertain in my mind, the actual moment of our meeting is crystal clear.

We met at lunch in the Princeton Cafeteria at the long table where the more boisterous and loud young post-docs tended to congregate. Suddenly Chern was in our midst and immediately he seemed as much at home with us, as we were with him. Like all great spirits, Chern treats people equally; the high and mighty can expect no courtesy from him that he would not also naturally extend to the lowliest among us.

On my side, at least, it was love at first sight. The beautiful mathematics that poured out of him, the love of geometry that was tangibly present in all his lectures, combined with a certain Chinese world view, made him irresistible.

In those early days my generation of geometers would watch Chern at the board with a sense of awe. He was a magician. How can one take a point p on M, and then calmly manipulate with dp? It took us years to master this trick! And the moment we understood one mysterious maneuver he would come up with the next one. And if one protested, he would agree with the objection, but calmly continue with what he was doing. And so, slowly, he got us all started on the road of understanding geometry in the Cartan-Chern tradition.

It was my good fortune to have Chern as a companion – as well as mentor – on many trips to exotic parts of the world; culminating with the first International conference in China where he served as guide, host, as well as flunky. Indeed, it turned out that Chern had personally stood in line for hours for our tickets to enter China via Hong Kong. Of course the day after he introduced us to the Vice Premier of the place. During all these global excursions I loved to steal a little time from mathematics to discuss other aspects of life and was always delighted and refreshed by the gentle good humor in which Chern would parry my often impertinent thrusts with a much more Eastern point of view than a Hungarian could possibly muster.

Eventually, during a summer's stay at Stanford we started to collaborate. Our hope was to redo the equidistribution theory of Nevanlinna and Ahlfors in a more geometric and up-to-date setting. Our efforts were not quite as successful as we had hoped in this direction; however, the refinement of "holomorphic Chern classes" which we formulated there has been taken up with great enthusiasm and vastly extended by the modern workers in Arakelov theory. By now it has progressed far beyond our understanding.

In any case, this collaboration was a joy from beginning to end. Compared with my other - also wonderful - collaborations, with Atiyah, Segal, Haefliger say - I suddenly found myself the more functorial partner dealing with someone who could compute circles around me. And, as is usually the case in such situations, the more functorial partner gets to write the final version. But Chern's explicit computations were too lovely to be cast away and so they appear in a companion paper. In fact, quite recently Chern told me that he has now seen how to extend and simplify these calculations significantly.

Chern is about ten years older than I am, and so I have always looked to him and followed his wonderful career as a sort of absolute upper bound of what the future might hold. It is a great inspiration to all of us, then, to see him continue steadfastly in his own work and in his devotion to our muse.

Thank you, dear friend. Please do not change, and please do not cease to be our guide for years and years to come.

S.S. Chern as Teacher

Louis Auslander
Grad. School and Univ. Ctr.
CUNY

I studied with S.S. Chern at the University of Chicago for the academic years 1951-1953.

This sentence might have begun "Professor S.S. Chern was my teacher" or "I was a student of Professor Chern", but neither of those phrases would have conveyed my feeling of a shared experience nearly as well. The sentence might have continued "during" instead of "for the academic years", but that would not have conveyed the complete dedication of the experience. My problem in writing this essay is how to find the right words to describe the experience, because, after all these years, the experience itself still seems to have happened only a short time ago.

I arrived in Chicago in the Fall of '51, found a place to live for my wife and myself, and went off to register. In those days, registration at The University of Chicago took place in the gym. Behind a table sat the adviser for the new mathematics students. We chatted a bit. I said I was interested in topology and had a course in differential geometry and a course in algebraic topology at Indiana University. The gentleman behind the table suggested I take Fiber Bundles from A. Weil, Riemannian Geometry from S.S. Chern, Lie Groups from E. Spanier and Algebraic Topology from S. MacLane. I naively agreed. I attended my first class in Riemannian Geometry and discovered that the gentleman behind the table in the gym was Chern and that his course really had as a prerequisite a course in manifold theory that he had taught the previous quarter. That quarter was my first and, perhaps, my most violent encounter I was to have with one of Chern's axioms of faith – mathematicians learn mathematics all the time.

Again we come to one of the crucial words in this narrative. One might try "a mathematician studies mathematics," but this would be too passive to convey Chern's belief. Chern believed that mathematicians do mathematics and part of doing mathematics is absorbing mathematics. He also believed that mathematicians do mathematics awake and asleep.

In the Spring of '52 I took the written qualifying examination. In those days this consisted of four three hour examinations in algebra, geometry-topology, real analysis and complex analysis. Before taking this exam I had asked Chern to be my thesis adviser and he had agreed. After the examinations were graded, Professor Barnard, the student faculty adviser, called me into his office and said that I had done very well in all the parts of the examination except the geometry part and did I really want to do my thesis in geometry. Professor Barnard, I am now certain, was trying to be kind. I took silent exception to his approach, never explaining that my wife had recently arrived home with a newborn son and had hemorrhaged the night before the geometry examination. However, when I saw Chern and asked him if he still wanted be my thesis adviser, he conveyed the understanding that examinations were not important – it was now the time to do mathematics.

This then began a process of education, an apprenticeship, by indirection. Chern would say things like: "Would you look at Finsler geometry?"; or, "It would be very nice if we meet in my office one day a week and talk things over." No matter what I presented, Chern would listen politely and almost silently. On occasion he would say, "I do not understand." I soon learned that "I do not understand" was a euphemism for "That's wrong!!"

Somehow Chern conveyed the philosophy that making mistakes was normal and that passing from mistake to mistake to truth was the doing of mathematics. And somehow he also conveyed the understanding that once one began doing mathematics it would naturally flow on and on. Doing mathematics would become like a stream pushing one on and on. If one was a mathemtician, one lived mathematics ... and so it has turned out.

Reminiscences and Acknowledgements

Haruo Suzuki
Department of Mathematics
Hokkaido University

The word "dàrén (大 人)" seems to be the word for Chern. According to the dictionary, "dàrén" has some meanings; a great man, a large hearted man, a great scholar and also a giant. Yes, he is. "Mèngzǐ (孟 子)", a well known Chinese thinker said that "dàrén bùshī chìzǐxīn (大人不失赤子心)". It has two meanings: Firstly, "dàrén" keeps a very pure mind like baby. Secondly, "dàrén" tries to understand his people very well and always wish their happiness. Chern is very such a man.

It was 36 years ago – the end of September, 1955 that I met S.-S. Chern for the first time. I went to the University of Chicago to study geometry. My teacher in Tohoku University, Japan, S. Sasaki who was a friend of Chern recommended me to go to Chicago.

At that time I was working at homotopy theory but I was very much interested in the theory of fiber bundles and characteristic classes by which Chern succeeded in getting a beautiful proof of Gauss-Bonnet formula for closed Riemannian manifolds.

Chern had advanced courses in geometry; Complex manifolds I,

II in the Autumn Quarter, 1955 and the Winter Quarter, 1956. His convincing, powerful voice remains in my ears. They began by Kähler geometry then moved to sheaves, connections, characteristic classes and concluded by the Riemann-Roch theorem for algebraic varieties. My knowledge obtained from these courses is still now useful.

Chern conducted two seminars. The one was a seminar on topology and geometry, cooperated by E.H. Spanier. The other was a seminar on complex manifolds. In the former, I could see E.H. Brown, M. Hirsch, R. Lashof and F.P. Peterson. Sometimes S. Smale appeared. In the first day of the seminar in a Quarter, we had an organization meeting and then speakers of all weeks were decided. A talk which I remember was that given by Lashof on the total curvature of immersed manifold (a joint work with Chern).

In the seminar on complex manifolds, I could see R. Gunning, H. Levine, T. Matsusaka and B. Reinhart among members. The office of A. Weil was in the opposite side to the office of Chern across a corridor. They were very friendly, so I could have chance to talk to Weil sometimes.

In lunch time and coffee time, many mathematicians in Eckhart Hall gathered into one table in Hutchinson Commons or a tiffin room, we talked mathematics without rigorous arguments. When S. Sternberg visited Chicago, Chern joined us in the tiffin room to coffee.

During about one month in summer, 1956, an international symposium of algebraic topology was held in the National University of Mexico. From Chicago, Chern and many members of his seminars participated in it. Among them I was. Speakers in the symposium were mathematicians of highest level with topology and geometry in the world at that time, including Chern. I was very much excited to listen their talks. In an excursion, I took a picture of Chern, S. Eilenberg and F. Hirzebruch together.

In that meeting, Chern introduced R. Thom to me. Thom had made good works on realizations of cohomology classes by submanifolds which I was interested in. I appreciate Chern to have given me a chance to meet Thom.

In the Autumn Quarter, 1956, Gunning and Peterson left Chicago for Princeton. Instead, M. Kuranishi came, he was working at involutive differential systems. Chern interested in his results. I could get some knowledges on that topic.

Chern showed me W.-T. Wu's papers on Pontrjagin classes I, II, III, which were written in Chinese. I could not speak Chinese but read it to some level. I reported results of my studies on these papers in his seminar. They were quite useful in my thesis work later.

During Summer Quarter, 1957, I finished my thesis under advice by Chern. In this quarter, the Summer Institute of Geometry was held in Eckhart Hall. I could see again R. Thom who was helpful to my works on the realization of the Stiefel-Whitney classes.

One day, I happened to meet Chern on South 57th Street and we walked to Eckhart Hall together. He asked me why I was interested in the realization of Stiefel-Whitney classes. My answer was that Chern got realization of characteristic classes by algebraic subvarieties. I got

Ph. D. degree in the Rockefeller Memorial Chapel, August 30, 1957 together with H. Levine and D. Hertzig.

Chern moved to the University of California, Berkeley 1960. In summer 1979, Chern Symposium was held there in memory of his emeritus position. I joined it and got a chance to say congratulations to him.

I admitted to be a member of MSRI founded by Chern, for the special year of symplectic geometry 1988–89. Actually, I stayed there twice, August–September, 1988 and May–June, 1989. I could accomplish my work on holonomy groupoids of foliations with singularities such as those of symplectic leaves of Poisson manifolds. A. Weinstein who got Ph. D. under Chern, was the Chairman of the program of symplectic geometry in MSRI.

In May–June 1989, Chern appeared almost every day in MSRI. One day he invited me together with T. Ozawa to a lunch in Mandarin Garden (a Chinese Restaurant in Berkeley). He explained his deep thought on geometry. He emphasized that the concept of geometry is developing gradually in these days. Besides, then I could hear a part of his view on state of things in the world. These conversations were quite instructive to me.

Chern visited Japan several times. He visited the University of Tokyo in about winter 1960 and then Kyoto University to participate in the U.S.-Japan Seminar 1965. When he was in Tokyo to attend a conference on differential geometry 1974, Y. Katsurada and I invited him to Hokkaido University, Sapporo. The title of his talk there was "Real surfaces in complex manifolds". We enjoyed his visiting very much.

In October 1987, he visited Japan to give a talk on "Dupin submanifolds" in the Differential Geometry Symposium held in Tohoku University, Sendai. Then I was told by him that I could apply to MSRI for the year 1988–89. I met Chern about 10 times so far. In any times, I got some good things from conversations with him and was encouraged to work at mathematics.

I am deeply grateful to him. I wish him good health and a continuations of excellent mathematics.

S.S. Chern: Always changing, always the same.

Phillip A. Griffiths
Director
Institute for Advanced Study

S.S. Chern came into my life more than thirty years ago. While a first year graduate student at Princeton, I had become interested in what is now called global differential geometry. As a student of Don Spencer, I was also naturally interested in complex manifolds. Don had a copy of lecture notes on complex manifolds that Chern had given in Brazil, notes that took the subject from its first principles through the recently proved Hirzebruch-Riemann-Roch theorem. These notes had a cult following, similar to that enjoyed by Chern's "140" and "240" lecture notes from Berkeley.

Don suggested that I go to Berkeley for the summer. As anyone who has ever spent a summer in Princeton can appreciate, this was an opportunity which was difficult to refuse. So I rode the train out, found a room in a boarding house, and went over to the math building to see if I might make an appointment to talk to Chern. He was expecting me at some point, as Don had written a letter of introduction. That day he was in his office, and after a very warm greeting Chern invited me down the street for a lunch at a Chinese restaurant. Over the summer he was both my teacher and colleague, as he was to so many aspiring young mathematicians. Our discussions were relaxed but substantive, with the complementarity between the big picture and interesting special cases communicated by example. Mathematics was set in historical perspective, yet Chern was always looking ahead to what the future interesting problems would be. It is difficult to adequately express in words the effect that this experience had on me, both mathematically and otherwise.

Over the more than thirty intervening years, Chern and I have remained close friends and colleagues. A visit to Berkeley always brings with it a mathematical discussion, a meal – usually Chinese and always excellent, and reflections on the general state of the world. During this period I have observed my own experience being replicated time and again. Many mathematicians at all levels, but especially young ones, have enjoyed the same fruitful relationship, mathematical reflections, encouragement, friendship and Chinese food that have made such a difference to me personally.

Chern is genuinely interested in the work and ideas of students just finding their way. He is encouraging yet is willing to say some idea may not be interesting. He demonstrates a combination of wisdom, mathematical discrimination and tact. He always treats one with respect, as a colleague and equal. In addition to the mathematical relationship, he shows a real interest in the person in a broader sense, asking about his family, career plans, and travel, discussing world politics, history and events with as much wisdom as he shows in mathematical discussions.

Chinese food is always part of a relationship with Chern and Mrs. Chern; I cannot count the number of outstanding Chinese meals my wife and I have enjoyed thanks to the Chern's generosity.

Finally, Chern has provided me and many other mathematicians with important contacts all over the globe. He has enabled us to travel to China and introduced us to many fine people who have become close friends.

Long before the concept of "mentoring" came into vogue Chern was a model mentor. For those just embarking on a career in mathematics, as I was those thirty-odd years ago, the experience described above can be decisive. A beginning student needs to learn more than facts and techniques: he or she needs to absorb a sort of world view of mathematics, a set of criteria with which to judge whether or not a problem is interesting, a method of passing on mathematical knowledge and enthusiasm and taste to others. To most fully develop as a mathematician one needs a mentor who can provide what Chern has provided for so many: formal teaching, teaching by example, encouragement, realism, and contacts.

The chain assures good teaching and mentoring for the future as well. Because of Chern, Don Spencer and others, I was exposed to a concept of mentoring and hope to have passed this on. Of my former students there are many who I believe are fine teachers and mentors; nearly all have had the opportunity to meet and work with Chern.

Much has been written recently about a developing crisis in scientific and mathematical education. There are a number of reasons for this. From early in undergraduate college, science and math courses are traditionally used to "filter out" students; there is an assumption that the truly good students will rise to the top in spite of the dryness and competitiveness of introductory courses. While there are undoubtedly students who are sufficiently single-minded and self-motivated to survive this system, many good and even excellent students are now diverted by the system into other fields. As a community we need to develop the tradition of recruiting students, and to move away from our old way of just selecting them.

Math and science cannot be successfully taught simply as a set of facts to be memorized or a set of techniques to be mastered. One needs to develop a relationship to the subject. There is a way of seeing, a set of criteria by which to judge a problem, an emotional and aesthetic as well as an intellectual aspect that one can best learn from someone like Chern. Those who have not experienced this side of the subject have difficulty projecting to their students the excitement that can come from understanding a really beautiful solution to a problem.

Chern is retired now for some years. However, through his former students, colleagues, friends and others his influence is enormous. We all hope it will remain so.

I want personally express to Chern my profound gratitude for his generosity and friendship during these past thirty-plus years.

Shiing–Shen Chern's Influence on Value Distribution Dedicated to Shiing–Shen Chern

Wilhelm Stoll*

University of Notre Dame

Department of Mathematics

Shiing–Shen Chern is one of the great mathematicians of our time. His work stretches all over differential geometry and into many related areas. He created new foundations and erected novel structures for the present and for the future. He renovated my own field, value distribution theory of several complex variables with differential geometric ideas. I am deeply honored by the request to reflect upon his achievements and upon our relationship. This task is not easy. I am not a differential geometer and I leave it to others to describe his impact upon this field and upon mathematics altogether. Perhaps, I can shed some light upon his influence on the growth of value distribution theory. Although we have known one another for a long time, we only met at various isolated occasions. Never were we colleagues at the same institution nor did we collaborate on a joint mathematical venture. Yet he deeply influenced my life and my work. Memories fade in the fog of

*Partially supported by National Science Foundation Grant DMS 91-00872

the past. Thus I can report only as I remember and how I remember these things.

I do not recall when we met first. In 1954/55, he talked in the colloquium of the University of Pennsylvania, Philadelphia, a risky undertaking in those days. For another speaker, Besicovitch, the department faculty had to collect enough money among themselves to buy him a return ticket. Fortunately, Chern encountered no such difficulty. We talked about my extension [24] of Nevanlinna's two main theorems to meromorphic maps $f : M \to \mathbb{P}_n$. Here M is a non-compact, connected complex manifold endowed with a positive, closed C^∞-form χ of bidegree $(m-1, m-1)$, for instance a Kaehler manifold. Of course $\mathbb{P}_n = \mathbb{P}(\mathbb{C}^{n+1})$ is the n-dimensional complex projective space. I combined the work of Hellmuth Kneser [19] with the work of Lars Ahlfors [1], and Hermann Weyl and Joachim Weyl [38]. Chern urged me to rework the paper with the infusion of differential geometry. Deeper insights would be gained and more people would be attracted. At that time, I was distracted by other problems, blowing up points, defining meromorphic maps and extending analytic sets. Thus the suggestion had to wait and Chern himself carried differential geometry into value distribution theory.

Around this time, we also met at a weekend conference at Columbia University. I do not remember much, except that Chern assembled a group of us and took us to a small Chinese restaurant nearby which served an excellent, plentiful lunch for 50 cents.

In 1955, I returned to Tübingen, Germany. From August 1957 to April 1959 my family and I were at the Institute for Advanced Study in Princeton. In 1958, a Summer School was held at the University of Chicago. Upon my inquiry, Chern invited me. Seemingly I achieved little during this month in Chicago. However, the questions which I was asked there initiated a long line of research. Chern gave me the task to integrate the "Unintegrated First Main Theorem" of Harold

Levine [21], a student of Chern. The target family was the set of all k-codimensional complex projective, linear subspaces of $\mathbb{P}_n$. An integration method was invented by the Weyls [38] on Riemann surfaces which extended to complex manifolds M of dimension $m > 1$ ([24]). Thus we would assume $m \geq k$ and the existence of a closed, positive form χ of bidegree $(m-k, m-k)$ and of class C^∞ on M. Let G and g be open subsets of M such that $\bar{G}$ is compact and $\bar{g} \subset G$. Moreover assume that the boundaries ∂G and ∂g are smooth and of pure dimension $2m-1$. Then the Dirichlet problem $dd^c\psi \wedge \chi \equiv 0$ on $G - \bar{g}$ would have to be solved with $\psi|\partial g = 0$ and $\psi|\partial G = R > 0$ constant. If $k = 1$, there is always a unique solution. If $k \geq 2$, there is no solution in general. Thus the condensor method of the Weyls fails if $k > 1$. I could not think of any other method to obtain the classical First Main Theorem in Levine's case. Therefore I told Chern, that it was impossible to "integrate" Levine's Theorem. Shortly afterwards Shiing–Shen Chern [5] integrated Levine's Theorem in the case $m = n = k$. I was very much surprised indeed. How did Chern do it? Simply he refused to solve the unsolvable and picked any ψ satisfying the boundary condition! As compensation he had to admit another new term to the formula of the First Main Theorem. There were several advantages. Because $m = k$, he can take $\chi = 1$. For ψ he takes an exhaustion. Since he took $M = \mathbb{C}^m$, the exhaustion $\psi(z) = \|z\|^2$ is convenient and yields the Kaehler from $dd^c\psi$ in the new term. Since $f : \mathbb{C}^m \to \mathbb{P}_m$ is assumed to be an open, holomorphic map the fibers of f consist of isolated points. They are easy to count and for a particular fiber g and G can be taken such that the fiber does not intersect the boundaries. Also the projective linear planes of codimension m in $\mathbb{P}_m$ are exactly the points of $\mathbb{P}_m$. Thus he arrives at a condition which assures that "f assumes almost every value" that is $\mathbb{P}_m - f(\mathbb{C}^m)$ has measure zero.

In Chicago I had forgotten about an admonition of Max Planck:

"The energy principle is not a law of nature but of man. Each time it fails in nature, we invent a new type of energy which balances the equation." The final resolution of Chern's request in Chicago requires that the theorems of Levine and Chern are proved in the general situation without all the simplifying assumptions, which was carried out in [25], [26], [27] and [28]. Let me outline the answer here.

Let V be a complex vector space of dimension $n+1>1$. Put $V_* = V - \{0\}$. Then $\mathbb{C}_* = \mathbb{C} - \{0\}$ acts on V_* by multiplication. The quotient space $\mathbb{P}(V) = V_*/\mathbb{C}_*$ is called the *complex projective space* and trivially is a connected, compact, complex manifold of dimension n. The quotient map $\mathbb{P} : V_* \to \mathbb{P}(V)$ is open and holomorphic. If $S \subseteq V$, abbreviate $\mathbb{P}(S) := \mathbb{P}(S \cap V_*)$. Take $p \in \mathbb{Z}[0,n]$. The *Grassmann cone*

$$\widetilde{G}_p(V) = \{\mathfrak{a}_0 \wedge \cdots \wedge \mathfrak{a}_p | \mathfrak{a}_j \in V \text{ for } j = 1, \ldots, p\} \tag{1}$$

is an analytic subset of the exterior product $\bigwedge_{p+1} V$. The Grassmann manifold $G_p(V) = \mathbb{P}(\widetilde{G}_p(V))$ is a compact, connected, complex-submanifold of $\mathbb{P}(\bigwedge_{p+1} V)$ with dimension $d(p,n)$ and degree $D(p,n)$:

$$d(p,n) = (n-p)(p-1) \qquad D(p,n) = d(p,n)! \prod_{\lambda=0}^{p} \frac{\lambda!}{(p-\lambda)!} \cdot \tag{2}$$

Take $a \in G_p(V)$. Then $a = \mathbb{P}(\mathfrak{a})$ where $\mathfrak{a} = \mathfrak{a}_0 \wedge \cdots \wedge \mathfrak{a}_p \neq 0$ and $\mathfrak{a}_j \in V$ for $j = 0, \ldots, p$. Then $E(a) = \mathbb{C}\mathfrak{a}_o + \cdots + \mathbb{C}\mathfrak{a}_p$ is a linear subspace of dimension $p+1$ of V, and $\ddot{E}(a) = \mathbb{P}(E(a))$ is a p-dimensional projective linear subspace of $\mathbb{P}(V)$. If $x = \mathbb{P}(\mathfrak{x}) \in \mathbb{P}(V) - \ddot{E}(a)$, then $\mathfrak{x} \wedge \mathfrak{a} \neq 0$ and $x \wedge a = \mathbb{P}(\mathfrak{x} \wedge \mathfrak{a})$ is well defined.

Select a hermitian metric $(\cdot|\cdot)$ on V. It induces a hermitian metric on $\mathbb{P}\left(\bigwedge_{p+1} V\right)$. The norm is defined by $\|\mathfrak{x}\| = \sqrt{(\mathfrak{x}|\mathfrak{x})}$. A strictly parabolic exhaustion $\tau = \tau_V$ on V is defined by $\tau(\mathfrak{x}) = \|\mathfrak{x}\|^2$. The *Fubini-Study* from $\Omega = \Omega_V$ on $\mathbb{P}(V)$ is uniquely defined by $\mathbb{P}^*(\Omega) = dd^c \log \tau$. Here $d^c = (i/4\pi)(\overline{\partial} - \partial)$. The form Ω is positive, real analytic and has bidegree (1,1). The degree of pure q-dimensional analytic subset A of $\mathbb{P}(V)$ is given by

$$\deg A = \int_A \Omega^q . \tag{3}$$

Abbreviate $\Omega_p = \Omega_{\underset{p+1}{\wedge} V}$. Then

$$D(p,n) = \int_{G_p(V)} \Omega_p^{d(p,n)} . \tag{4}$$

If $x = \mathbb{P}(\mathfrak{x}) \in \mathbb{P}(V)$ and $a = \mathbb{P}(\mathfrak{a}) \in G_p(V)$ define the *distance* from x to $\ddot{E}(a)$ by

$$0 \leq \Box x \dot{\wedge} a \Box = \frac{\|\mathfrak{x} \dot{\wedge} \mathfrak{a}\|}{\|\mathfrak{x}\| \ \|\mathfrak{a}\|} \leq 1 . \tag{5}$$

If $a \in G_{n-1}(V)$ and $x \in \mathbb{P}(V) - \ddot{E}(a)$, then $\Omega(x) = -dd^c \log \Box x \dot{\wedge} a \Box^2$.

A connected, compact complex submanifold F_p of dimension $n + p(n-p)$ of $\mathbb{P}(V) \times G_p(V)$ is defined by

$$F_p = \{(x,a) \in \mathbb{P}(V) \times G_p(V) | x \in \ddot{E}(a)\} . \tag{6}$$

The projections $\rho : F_p \to G_p(V)$ and $\pi : F_p \to \mathbb{P}(V)$ have pure fiber dimensions p and $p(n-p)$ respectively. If $a \in G_p(V)$ then $\rho^{-1}(a) = \ddot{E}(a) \times \{a\}$ and $\pi(\rho^{-1}(a)) = \ddot{E}(a)$.

Let M be a connected, complex manifold of dimension m. Define $k = n - p$ and $q = m - k$. Assume that $q \geq 0$ and $k \geq 1$. Let $f : M \to \mathbb{P}(V)$ be a holomorphic map. The pull back of $\pi : F_p \to \mathbb{P}(V)$ by f is a connected, complex manifold

$$F_p(f) = \{(x,a) \in M \times G_p(V) | f(x) \in \ddot{E}(a)\} \tag{7}$$

of dimension $m+(p+1)k$. Let $\pi_f : F_p(f) \to M$ and $\rho_f : F_p(f) \to G_p(V)$ be the projections. Then π_f has pure fiber dimension pk. If $a \in G_p(V)$, then $\rho_f^{-1}(a) = f^{-1}(\ddot{E}(a)) \times \{a\}$ and $S_a := f^{-1}(\ddot{E}(a)) = \pi_f(\rho^{-1}(a))$. Here $\dim_x S_a \geq q$, if $x \in S_a$. Here f is said to be *general of order* k *for* $a \in G_p(V)$, if either $S_a = \emptyset$ or $\dim_x S_a = q$ for all $x \in S_a$. If f

is general of order k for all $a \in G_p(V)$, then f *is general of order* k which is the case if and only if ρ_f is an open map. For simplicity we shall assume that f is general of order k. The following diagram (8) commutes.

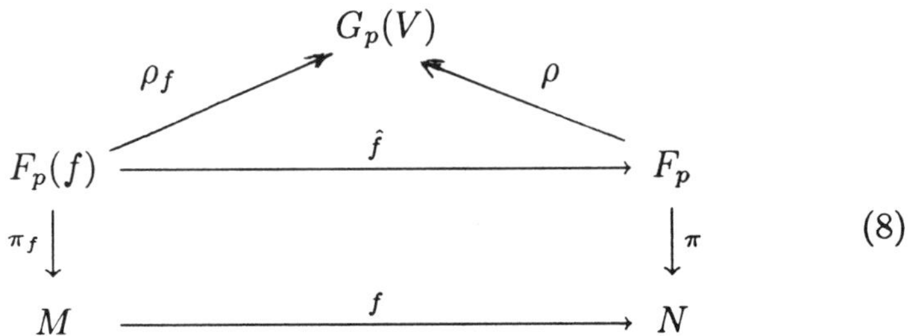

The map ρ_f has pure fiber dimension q. By [25], the mapping multiplicity $\nu_{\rho_f}(x,a) \geq 0$ is defined for all $(x,a) \in F_p(f)$. Take any $x \in M$ and $a \in G_p(V)$. If $x \in S_a$, then $(x,a) \in F_p(f)$; define $\nu_{f,a}(x) = \nu_{\rho_f}(x,a) > 0$. If $x \in M - S_a$ put $\nu_{f,a}(x) = 0$. Then $\nu_{f,a}(x)$ is called the *intersection multiplicity* of f at x with $\ddot{E}(a)$.

If $q = 0$, the *counting function* $n_{f,a}$ of f for $a \in G_p(V)$ is defined for any compact subset $K \neq \emptyset$ of M by

$$n_{f,a}(K) = \sum_{x \in K} \nu_{f,a}(x) < \infty \,. \tag{9}$$

If $q > 0$, we assume that there is given a closed, non-negative form χ of bidegree (q,q) and class C^∞ on M. For each compact subset K of M, abbreviate $K_a = K \cap S_a$ and define the *counting function* $n_{f,a}$ of f for $a \in G_p(V)$ by

$$n_{f,a}(K) = \int_{K_a} \nu_{f,a}\chi < \infty \,. \tag{10}$$

Both notations can be combined by taking $\chi = 1$ and $\int_{K_a} = \sum_{x \in K_a}$ if $q = 0$. The *spherical image function* $A_f = A_{k,f}$ of f for order k is defined for each compact subset K of M by

$$A_f(K) = \int_K f^*(\Omega^k) \wedge \chi \geq 0 \,. \tag{11}$$

If M is compact and $a \in G_p(V)$, then $A_f(M) = n_{f,a}(M)$. If M is not compact, we take a relative compact, open subset H of M with smooth boundary. Then we will need a boundary integral to save the identity. Some preparations are needed to define the integrand.

Take $a \in G_p(V)$ and $\mathfrak{a} \in \widetilde{G}_p(V)$ with $\mathbb{P}(\mathfrak{a}) = a$ and $\|\mathfrak{a}\| = 1$. A linear map $\alpha_{\mathfrak{a}} : V \to \bigwedge_{p+2} V$ is defined by $\alpha_{\mathfrak{a}}(\mathfrak{x}) = \mathfrak{x} \wedge \mathfrak{a}$. Then $E(a)$ is the kernel of $\alpha_{\mathfrak{a}}$. A holomorphic map $\ddot{\alpha}_{\mathfrak{a}} : \mathbb{P}(V) - \ddot{E}(a) \to \mathbb{P}\left(\bigwedge_{p+2} V\right)$ is defined by $\ddot{\alpha}_{\mathfrak{a}} \circ \mathbb{P} = \mathbb{P} \circ \alpha_{\mathfrak{a}}$. On $\mathbb{P}(V) - \ddot{E}(a)$ define

$$\Phi(a) = (\ddot{\alpha}_{\mathfrak{a}})^*(\Omega_{p+1}) \geq 0 \,. \tag{12}$$

There is one and only one $b \in G_{n-p-1}(V)$ such that

$$E(b) = \{\mathfrak{x} \in V | (\mathfrak{x}|\mathfrak{z}) = 0 \quad \text{for all} \quad \mathfrak{z} \in E(a)\} = E(a)^{\perp} \,. \tag{13}$$

Then $V = E(a) \oplus E(b)$. Let $\beta_a : V \to E(b)$ be the projection with $\ker \beta_a = E(a)$. Thus there is an injective linear map $\gamma_a : E(b) \to \bigwedge_{p+2} V$ such that $\gamma_{\mathfrak{a}} \circ \beta_a = \alpha_{\mathfrak{a}}$. Here $\gamma_{\mathfrak{a}}$ is an isometry.

$$\ddot{\beta}_{\alpha} : \mathbb{P}(V) - \ddot{E}(a) \to \mathbb{P}\big(E(b)\big) \quad \text{and} \quad \ddot{\gamma}_{\mathfrak{a}} : \mathbb{P}\big(E(b)\big) \to \mathbb{P}\Big(\bigwedge_{p+2} V\Big) \tag{14}$$

are defined by $\ddot{\beta}_a \circ \mathbb{P} = \mathbb{P} \circ \beta_a$ and $\ddot{\gamma}_{\mathfrak{a}} \circ \mathbb{P} = \mathbb{P} \circ \gamma_{\mathfrak{a}}$. Then $\ddot{\alpha}_{\mathfrak{a}} = \ddot{\gamma}_{\mathfrak{a}} \circ \ddot{\beta}_a$ and $\Omega_b = \gamma_a^*(\Omega_{p+1})$ is the Fubini Study form on $\ddot{E}(b)$. Hence

$$\Phi(a) = \ddot{\beta}_a^*(\Omega_b) \,. \tag{15}$$

Because $\dim \ddot{E}(b) = n - p - 1 = k - 1$, we obtain $\Phi(a)^k \equiv 0$. On $\mathbb{P}(V) - \ddot{E}(a)$ define the *Levine form* $\Lambda(a)$ by

$$\Lambda(a)(x) = -\log \Box x \dot{\wedge} a \Box^2 \sum_{\nu=0}^{k-1} \big(\Phi(a)(x)\big)^{\nu} \wedge \big(\Omega(x)\big)^{k-\nu-1} \,. \tag{16}$$

Since $dd^c \log \Box x \dot{\wedge} a \Box^2 = \Phi(a) - \Omega$, on $\mathbb{P}(V) - \ddot{E}(a)$ we obtain

$$dd^c \Lambda(a) = \Omega^k \,. \tag{17}$$

Let $H \neq \emptyset$ be an open subset of M such that $\overline{H}$ is compact. Assume that $\partial H = \overline{H} - H$ is a smooth boundary manifold of H. Let ψ be a function of class C^2 on M such that for each point $x \in \partial H$ there is an open neighborhood U of x such that either $\psi(z) \geq \psi(x)$ for all $z \in \overline{H} \cap U$ or $\psi(z) \leq \psi(x)$ for all $z \in \overline{H} \cap U$. Now [27] Proposition 4.3 yields

$$\int_{\partial H} f^*(\Lambda(a)) \wedge d^c\psi \wedge \chi = \int_H f^*(\Lambda(a)) \wedge dd^c\psi \wedge \chi + \int_H d\psi \wedge f^*(d^c\Lambda(a)) \wedge \chi \,. \tag{18}$$

Take $a \in G_p(V)$. Let K be the support of $\psi\chi$ on ∂H. Assume that $K \cap S_a$ has measure zero on the analytic set S_a. Theorem 4.4 in [27] implies

$$\begin{aligned} &\int_{\partial H} \psi f^*(d^c\Lambda(a)) \wedge \chi + \int_{H \cap S_a} \nu_{f,a}\psi\chi \\ &= \int_H d\psi \wedge f^*(d^c\Lambda(a)) \wedge \chi + \int_H \psi f^*(\Omega^k) \wedge \chi \,. \end{aligned} \tag{19}$$

Subtraction yields the *Green Residue Theorem* [27]

$$\begin{aligned} &\int_{\partial H} f^*(\Lambda(a)) \wedge d^c\psi \wedge \chi - \int_{\partial H} \psi f^*(d^c\Lambda(a)) \wedge \chi - \int_{H \cap S_a} \nu_{f,a}\psi\chi \\ = \; &\int_H f^*(\Lambda(a)) \wedge dd^c\psi \wedge \chi - \int_H \psi f^*(\Omega^k) \wedge \chi \,, \end{aligned} \tag{20}$$

which yields *First Main Theorems* by particular choices of ψ and χ. If we take $\psi \equiv 1$, we obtain the *Unintegrated First Main Theorem* [27]

$$A_{k,f}(\overline{H}) = n_{f,a}(\overline{H}) + \int_{\partial H} f^*(d^c\Lambda(a)) \wedge \chi \tag{21}$$

which was proved by Levine [21] in the case $q = 0$. Then S_a consists of isolated points and $\chi = 1$. Thus $K \cap S_a$ has measure zero on S_a if and only if $S_a \cap \partial H = \emptyset$. The identity (21) has the great disadvantage that the integrand of the boundary integral has unknown sign which could fluctuate. Integration is to eliminate this property.

For instance let u be a function of class C^∞ on M such that $u|\partial H = R > 0$ is constant and $u \leq R$ on H. Take $\psi = R - u$. Then $K \cap S_a = \emptyset$ for each $a \in G_p(V)$. Define

$$m_{f,a}(\partial H, u, \chi) = \int_{\partial H} f^*(\Lambda(a)) \wedge d^c u \wedge \chi \geq 0 \qquad (\textit{compensation function}) \tag{22}$$

$$N_{f,a}(H, u, \chi) = \int_{H \cap S_a} \nu_{f,a}(R - u)\chi \geq 0 \qquad (\textit{Valence function}) \tag{23}$$

$$T_f(H, u, \chi) = \int_H (R - u) f^*(\Omega^k) \wedge \chi \geq 0 \qquad (\textit{Characteristic function}) \tag{24}$$

$$\Delta_{f,a}(H, u, \chi) = \int_H (R - u) f^*(\Lambda(a)) \wedge dd^c u \wedge \chi \quad (\textit{Deficit}) \tag{25}$$

If H is *pseudoconvex*, i.e. if $dd^c u \geq 0$, then $\Delta_{f,a}(H, u, \chi) \geq 0$. Then (21) implies the *First Main theorem*

$$T_f(H, u, \chi) + \Delta_{f,a}(H, u, \chi) = m_{f,a}(H, u, \chi) + N_{f,a}(H, u, \chi) \, . \tag{26}$$

The deficit was introduced by Chern [5]. In order to eliminate it, we have to solve the Dirichlet problem $dd^c u \wedge \chi \equiv 0$ on H and $u|\partial U = R$. This problem is sensible only if $\chi > 0$ and $k = 1$, but in this case the unique solution is $u \equiv R$. Then (26) becomes the triviality $0+0 = 0+0$. This trap is avoided by a method of the Weyls [38]. Let G and g be open subsets of M such that $\overline{G}$ and $\bar{g}$ are compact with $\emptyset \neq \bar{g} \subset G$. Assume that $\partial G = \overline{G} - G$ and $\partial g = \bar{g} - g$ are smooth boundary manifolds of class C^∞ on $\overline{G} - g$ with $u|\partial G = R > 0$ and $u|\partial g = 0$ such that $0 \leq u \leq R$ on $H = G - \bar{g}$. Define $u \equiv 0$ on g and $u \equiv R$ on $M - \overline{G}$. On ∂H take the derivatives of u from inside H. Then $\Delta_{f,a}(H, u, \chi)$ is defined by (25), and $T_f(G, u, \chi)$ and $N_{f,a}(g, u, \chi)$ are defined by (24) and (23) replacing ∂H by ∂G respectively ∂g. There is a function $\tilde{u}$ of class C^∞ on M such that $\tilde{u}|\overline{H} = u$. Observe that $\tilde{u}|\partial G = R$ and

$\tilde{u}|\partial g = 0$. Apply (26) to G, $\tilde{u}$ and then to $g, \tilde{u}$ and subtract. Then the *First Main Theorem for condensors* is obtained

$$\begin{aligned} & T_f(G,u,\chi) + \Delta_{f,a}(G-\bar{g},u,\chi) \\ = & \, m_{f,a}(\partial G,u,\chi) - m_{f,a}(\partial g,u,\chi) + N_{f,a}(G,u,\chi) \, . \end{aligned} \tag{27}$$

If $k=1$, $\chi > 0$, then there exists one and only one continuous function u such that u is of Class C^∞ on $G - \bar{g}$ with $dd^c u \wedge \chi \equiv 0$ on $G - \bar{g}$ and $u|\bar{g} \equiv 0$ and $u|(M-G) \equiv R > 0$. In this case $\Delta_{f,a}(G-\bar{g},u,\chi) \equiv 0$. We obtain the First Main Theorem of Weyl [38] if $m=1$ and of [24] if $m > 1$.

$$T_f(G,u,\chi) = m_{f,a}(\partial G,u,\chi) - m_{f,a}(\partial g,u,\chi) + N_{f,a}(G,u,\chi) \, . \tag{28}$$

Exhausting M by the net of all $G \supset \bar{g}$ a defect relation was obtained in these papers if $k=1$ and $\chi > 0$.

The exhaustion by nets is very cumbersome. Thus an exhaustion function $u : M \to \mathbb{R}_+$ was used. If $S \subseteq M$ and $r \in \mathbb{R}_+$, define

$$S[r] = \{x \in S | u(x) \le r\} \, , \qquad S(r) = \{x \in S | u(x) < r\} \, , \tag{29}$$

$$S\langle r\rangle = \{x \in S | u(x) = r\} \, , \qquad S_* \; = \{x \in S | u(x) > 0\} \, . \tag{30}$$

We assume that u is continuous, unbounded on M and that $M[r]$ is compact for all $r \in \mathbb{R}_+$. We assume that u is of class C^∞ at least on M_*. Define

$$\mathfrak{E}_u = \{r \in \mathbb{R}^+ | du(x) \ne 0 \text{ for all } x \in M\langle r\rangle\} \, . \tag{31}$$

Then $\mathbb{R}^+ - \mathfrak{E}_u$ has measure zero. If $r \in \mathfrak{E}_u$, then $M[r] = \overline{M(r)}$ and $\partial M(r) = M\langle r\rangle$ is a boundary manifold of class C^∞ of $M(r)$.

If u is of class C^∞ on M and if $dd^c u \ge 0$ on M, then (M,u) is called a *pseudoconvex manifold*. If even $dd^c u > 0$ on M, then M is a

Stein manifold and (M, u) is said to be *strictly pseudoconvex*. In either case, define

$$m_{f,a}(r) = m_{f,a}(r,u) = m_{f,a}\big(M\langle r\rangle, u, (dd^c u)^q\big) \geq 0 \tag{32}$$

$$N_{f,a}(r) = N_{f,a}(r,u) = N_{f,a}\big(M(r), u, (dd^c u)^q\big) \geq 0 \tag{33}$$

$$T_{k,f}(r) = T_{k,f}(r,u) = T_f\big(M(r), u, (dd^c u)^q\big) \geq 0 \tag{34}$$

$$\Delta_{f,a}(r) = \Delta_{f,a}(r,u) = \Delta_{f,a}\big(M(r), u, (dd^c u)^q\big) \geq 0 \tag{35}$$

and we obtain the *Integrated First Main Theorem*

$$T_f(r) = N_{f,a}(r) + m_{f,a}(r) - \Delta_{f,a}(r)\ . \tag{36}$$

Also define

$$A_{k,f}(r) = \int_{M[r]} f^*(\Omega^k) \wedge (dd^c u)^q \geq 0 \tag{37}$$

$$n_{f,a}(r) = \int_{M[r]\cap S_a} \nu_{f,a} (dd^c u)^{q+1} \geq 0 \tag{38}$$

Then

$$\begin{aligned} T_{k,f}(r) &= \int_{M[r]} (r-u) f^*(\Omega^k) \wedge (dd^c u)^q \geq 0 \\ &= \int_{M[r]} \Big(\int_u^r dt\Big) f^*(\Omega^k) \wedge (dd^c u)^q \\ &= \int_0^r \int_{M[t]} f^*(\Omega^k) \wedge (dd^c u)^q = \int_0^r A_{k,f}(t)dt\ . \end{aligned} \tag{39}$$

Similar we obtain

$$N_{f,a}(r) = \int_0^r n_{f,a}(t)dt\ . \tag{40}$$

The classical definitions have dt/t, because classical value distribution uses another exhaustion. We will introduce this exhaustion now.

Let M be a connected, complex manifold of dimension m. Let $\tau : M \to \mathbb{R}^+$ be a function of class C^∞ on M such that $\tau_0 = \sqrt{\tau}$ is an exhaustion of M. Abbreviate

$$\upsilon = dd^c\tau \qquad \omega = dd^c \log\tau \qquad \sigma_q = d^c \log\tau \wedge \omega^q\ . \tag{41}$$

Assume that $\omega \geq 0$ on M_*. Then $\upsilon \geq 0$ on M. Also $d\sigma_q = \omega^{q+1}$. Assume that

$$\upsilon^m \not\equiv 0 \equiv \omega^m = d\sigma_{m-1} \,. \tag{42}$$

Then (M, τ) is said to be a parabolic manifold. If so, there is a positive constant $\varsigma > 0$ such that

$$\varsigma = \int_{M\langle r\rangle} \sigma_{m-1} \qquad \text{for all} \quad r \in \mathfrak{E}_\tau \tag{43}$$

$$\int_{M[r]} \upsilon^m = \varsigma r^{2m} > 0 \qquad \text{for all} \quad r \in \mathbb{R}^+ \,. \tag{44}$$

If $p \in \mathbb{N}[1, m]$ we have

$$\tau^{p+1}\omega^p = \tau\upsilon^p - p d\tau \wedge d^c\tau \wedge \upsilon^{p-1} \tag{45}$$

$$\upsilon^m = m\tau^{m-1} d\tau \wedge \sigma_{m-1} \,. \tag{46}$$

The parabolic manifold (M, τ) is said to be *strict* if $\upsilon > 0$ on M. By [34] the parabolic manifold (M, τ) is strict if and only if M has the structure of a hermitian complex vector space of dimension m such that $\tau(x) = \|x\|^2$ for all $x \in M$.

Take $0 < s < r$ and apply the condensor method to $G = M(r)$ and $g = M(s)$ and $\chi = \upsilon^q$. Put $H = G - \bar{g}$. If $q > 0$, define u by

$$u|H = \frac{1}{2q}\left(\frac{1}{s^{2q}} - \frac{1}{\tau^q}\right) \geq 0 \,, \tag{47}$$

where $u|M[s] = 0$ and $u|\big(M - M(r)\big) = R = \frac{1}{2q}\Big(\frac{1}{s^{2q}} - \frac{1}{r^{2q}}\Big)$. If $q = 0$, then $\chi = 1$ and

$$u|M[r] = \log^+ \frac{\tau}{s^2} \qquad u|\big(M - M(r)\big) = R = 2\log\frac{r}{s} \,. \tag{48}$$

Then

$$d^c u \wedge \chi = \frac{1}{2}\sigma_q \qquad dd^c u \wedge \chi = \frac{1}{2}\omega^{q+1} \text{ on } \overline{H} \,. \tag{49}$$

For $r > 0$, the spherical *image function* $A_{k,f}$ is defined by

$$A_{k,f}(r) = \frac{1}{r^{2q}} \int_{M[r]} f^*(\Omega^k) \wedge \upsilon^q , \tag{50}$$

Thus $A_{k,f}$ increase and $A_{k,f}(s) \to A_{k,f}(0) \leq 0$ for $r \to 0$. If $\upsilon > 0$ in a neighborhood of $M[0]$, then $A_{k,f}(0) = 0$ follows easily. Take $0 < s < r \in \mathfrak{E}_{\tau_0}$ with $s \in \mathfrak{E}_{\tau_0}$. Stokes Theorem implies

$$\begin{aligned} A_{k,f}(r) - A_{k,f}(s) &= \frac{1}{r^{2q}} \int_{M\langle r\rangle} f^*(\Omega^k) \wedge d^c\tau \wedge \upsilon^{q-1} \\ &\quad - \frac{1}{s^{2q}} \int_{M\langle s\rangle} f^*(\Omega^k) \wedge d^c\tau \wedge \upsilon^{q-1} \\ &= \int_{M\langle r\rangle} f^*(\Omega^k) \wedge d^c \log\tau \wedge \omega^{q-1} \\ &\quad - \int_{M\langle s\rangle} f^*(\Omega^k) \wedge d^c \log\tau \wedge \omega^{q-1} \\ &= \int_{M(r)-M[s]} f^*(\Omega^k) \wedge \omega^q \geq 0 . \end{aligned} \tag{51}$$

Thus $s \to 0$ implies the *Kneser identity* [19] (by continuity for all $r > 0$)

$$A_{k,f}(r) = \int_{M[r]} f^*(\Omega^k) \wedge \omega^q + A_{k,f}(0) . \tag{52}$$

Take $a \in G_p(V)$. For $r > 0$, the *counting function* $n_{f,a}$ is defined by

$$n_{f,a}(r) = \frac{1}{r^{2q}} \int_{S_a[r]} \nu_{f,a}\upsilon^q . \tag{53}$$

As above we obtain

$$n_{f,a}(r) - n_{f,a}(s) = \int_{S_a[r]-S_a[s]} \nu_{f,a}\omega^q \geq 0 , \tag{54}$$

if $0 < s < r$. Hence $n_{f,a}$ increases and the limit $n_{f,a}(0) = \lim_{s\to 0} n_{f,a}(s) \geq 0$ exists. Therefore

$$n_{f,a}(r) = \int_{S_a[r]} \nu_{f,a}\omega^q + n_{f,a}(s) . \tag{55}$$

For $0 < s < r$ we define the *Characteristic* $T_{k,f}$ by

$$T_{k,f}(r,s) = T_f\big(M(r), u, v^q\big) = \int_{M[r]} (R-u) f^*(\Omega^k) \wedge v^q \,.$$

If $q > 0$, then

$$\begin{aligned} T_{k,f}(r,s) &= \int_{M[r]-M[s]} \frac{1}{2q}\Big(\frac{1}{\tau^q} - \frac{1}{r^{2q}}\Big) f^*(\Omega^k) \wedge v^q \\ &\quad + \frac{1}{2q}\Big(\frac{1}{s^{2q}} - \frac{1}{r^{2q}}\Big) \int_{M[s]} f^*(\Omega^k) \wedge v^q \\ &= \int_{M[r]-M[s]} \int_{\tau_0}^{r} \frac{dt}{t^{2q-1}} f^*(\Omega^k) \wedge v^q \\ &\quad + \frac{1}{2q}\Big(\frac{1}{s^{2q}} - \frac{1}{r^{2q}}\Big) \int_{M[s]} f^*(\Omega^k) \wedge v^q \\ &= \int_s^r \int_{M[t]-M[s]} f^*(\Omega^k) \wedge v^q \frac{dt}{t^{2q-1}} \\ &\quad + \int_s^r \int_{M[s]} f^*(\Omega^k) \wedge v^q \frac{dt}{t^{2q-1}} \\ &= \int_s^r A_{k,f}(t) \frac{dt}{t} \,. \end{aligned} \tag{56}$$

If $q = 0$ we obtain the same result. Thus we have the classical form

$$T_{k,f}(r,s) = \int_s^t A_{k,f}(t) \frac{dt}{t} \geq 0 \,. \tag{57}$$

Since f is holomorphic, (47) holds for $s = 0$ as well. However that is false if f is meromorphic and the indeterminacy interects $M[0]$.

Similar the *valence function* is defined by

$$N_{f,a}(r,s) = N_{f,a}\big(M(r), u, v^q\big) = \int_s^r n_{f,a}(t) \frac{dt}{t} \geq 0 \,. \tag{58}$$

For $r \in \mathfrak{C}_{\tau_0}$, the *compensation function* $m_{f,a}$ is defined by

$$m_{f,a}(r) = m_{f,a}(M\langle r\rangle, u, v^q) = \frac{1}{2} \int_{M\langle r\rangle} f^*(\Lambda_a) \wedge \sigma_q \geq 0 \,. \tag{59}$$

If $0 < s < r$, then *deficit* $\Delta_{f,a}$ is defined by

$$\Delta_{f,a}(r,s) = \frac{1}{2} \int_{M[r]-M(s)} f^*(\Lambda_a) \wedge \omega^{q+1} \geq 0 \,. \tag{60}$$

Now (28) implies the *First Main Theorem on parabolic manifolds.*

$$T_{k,f}(r,s) = N_{f,a}(r,s) + m_{f,a}(r) - m_{f,a}(s) + \Delta_{f,a}(r,s) \tag{61}$$

for all $0 < s < r \in \mathfrak{E}_{\tau_0}$ with $s \in \mathfrak{E}_{\tau_0}$. The identify (51) extends $m_{f,a}$ to a continuous function on $\mathbb{R}^+$ such that (61) holds for all $0 < s < r$.

If $k = 1$, then $q = m - 1$ and $\omega^{q+1} = \omega^m \equiv 0$. Therefore $\Delta_{f,a}(r,s) \equiv 0$ and we obtain the *classical First Main Theorem*

$$T_f(r,s) = T_{1,f}(r,s) = N_{f,a}(r,s) + m_{f,a}(r) - m_{f,a}(s)\ . \tag{62}$$

Chern [5] derives a Casorati-Weirerstrass type theorem, which can be extended to all $k \in N[1,n]$. Recall (2) and (4). Let W be a finite dimensional complex vectorspace. Let $h : G_p(V) \to W$ be a vector function such that $h(\Omega_p)^{d(p,n)}$ is integrable. The *average* $\mathfrak{M}(h)$ of h is defined by

$$\mathfrak{M}(h) = \frac{1}{D(p,n)} \int_{G_p(V)} h(\Omega_h)^{d(p,n)} \in W\ . \tag{63}$$

Then $\mathfrak{M}(1) = 1$. Abbreviate

$$c(k,p) = \sum_{\nu=1}^{k} \sum_{\mu=0}^{p} \frac{1}{\nu + \mu}\ . \tag{64}$$

Then Theorem 2.11 of [27] states

$$\mathfrak{M}\big(\Lambda(a)\big) = c(k,p)\Omega^{k-1}\ . \tag{65}$$

Now (46) and (63) imply

$$\mathfrak{M}\big(m_{f,a}(r)\big) = \frac{1}{2} c(p,k) A_{k-1,f}(r) \tag{66}$$

$$\mathfrak{M}\big(\Delta_{f,a}(r,s)\big) = \frac{1}{2} c(p,k)\big(A_{k-1,f}(r) - A_{k-1,f}(s)\big)\ . \tag{67}$$

The First Main Theorem yields

$$T_{k,f}(r,s) = \mathfrak{M}\big(N_{f,a}(r,s)\big) > 0 \tag{68}$$

Thus $T_{k,f}(r,s) \to \infty$ for $r \to \infty$. For $r > 0$ define

$$\chi_r(a) = \begin{cases} 1 & \text{if } f(M[r]) \cap \ddot{E}(a) \neq \emptyset \\ 0 & \text{if } f(M[r]) \cap \ddot{E}(a) = \emptyset \end{cases} \qquad b_f(r) = \mathfrak{M}(\chi_r) \tag{69}$$

$$\chi(a) = \begin{cases} 1 & \text{if } f(M) \cap \ddot{E}(a) \neq \emptyset \\ 0 & \text{if } f(M) \cap \ddot{E}(a) = \emptyset \end{cases} \qquad b_f = \mathfrak{M}(\chi) \tag{70}$$

Then $0 \leq b_p(r) \leq 1$ and $0 \leq b_f \leq 1$ and $b_f(r) \to b_f$ for $r \to \infty$. Define

$$\delta_k(f) = \liminf_{r\to\infty} \frac{A_{k-1,f}(r)}{T_{k,f}(r,s)} \tag{71}$$

which does not depend on s. Take $0 < s < r$. The First Main Theorem states

$$N_{f,a}(r,s) - \chi_r(a)T_f(r,s) \leq \Delta_{f,a}(r,s) + m_{f,a}(s) \ . \tag{72}$$

Applying $\mathfrak{M}$, we obtain

$$\big(1 - b_f(r)\big)T_{k,f}(r,s) \leq \frac{1}{2}c(p,k)A_{k-1,f}(r) \ . \tag{73}$$

Thus

$$(1 - b_f) \leq \frac{1}{2}c(p,k)\delta_k(f) \ . \tag{74}$$

If $\delta_k(f) = 0$, then $f(M) \cap \ddot{E}(a) \neq \emptyset$ for almost all $a \in G_p(V)$. The estimate (73) permits some surprising observations. For instance, take $k = 1$ then $A_{0,f}(r) \equiv \varsigma > 0$ is constant. If M is strictly parabolic, then $M = \mathbb{C}^m$ and $\varsigma = 1$. We have $p = n - 1$. Take $n = 6$. If $T_f(r,s) > 123$, then $f\big(M(r)\big)$ intersects 99% of all hyperplanes in $\mathbb{P}_6$. The basic features of the outlined theory are due to Chern. He introduced the concept of deficit which opened up a new dimension in value distribution theory. He told me once to mention his trick (63) - (74) where ever possible. Thus I presented it here. Chern's method will give us other results stated below.

Other investigations originated during that month in Chicago. Andreotti participated in the Summer Institute. Andre Weil asked him some question, which resulted in joint work with Aldo and initiated enduring collaboration and friendship. In my second year at the Institute Aldo and I shared an office which helped much.

April 1959, my family and I returned to Tübingen, Germany. In January 1960 I learned that Chern was in Europe. With the authorization of Hellmuth Kneser I was able to invited him to lecture in the Tübingen Mathematics Colloquium. At the beginning of February 1960, Heinz Hopf invited me to speak in an International Colloquium on Differential Geometry and Topology at the E.T.H. in Zürich, June 1960. He warned me: Talks on Several Complex Variables are forbidden except for a big lecture by Chern on "Geometric Problems in Complex Analysis". At the conference I asked Hopf why he made this rule. He said, if you permit the people in complex analysis to speak, they will take over. The conference was a great success. Many nations were represented.

Chern's talk in Tübingen was arranged for the time after the conference. Together we took the train from Zürich to Tübingen. We discussed his new results in value distribution theory. Also he asked me if I would be interested in a position at the University of Notre Dame. After considerable reflection, I gave him some conditions which may induce me to accept such an offer. For a long time nothing happened and I did not think about it any more. Suddenly, one day in August, I received an overseas call from Professor Arnold Ross with Hans Zassenhaus standing by. They made an offer and expected me to accept on the spot. It did not go quite so quickly, but the day after Kennedy was elected, we arrived at Notre Dame. Thus Shiing-Shen Chern instigated a fundamental change in our life. Soon afterwards Chern moved from Chicago to Berkeley.

During the sixties we met on several occasions. I recall the remarkable summer institute in La Jolla, where he talked about holomorphic maps $f : M \to N$ between hermitian manifolds of the same dimension. In retrospect, it seems to me that the Jacobian sections and Theorem 14.6 in [31] may be perhaps of help in extending some results of Chern [9] to the case $\dim M \neq \dim N$. During the Academic year 1968/69 I was on leave from Notre Dame visiting Stanford. Chern urged me to go to Berkeley. Perhaps I should have taken his advice. I hoped to visit Berkeley often, but I underestimated the difficulty of commuting between the two places. Still, I saw him quite a number of times. Also he invited me to his house, where I could admire the breath taking view over the bay area.

The Stanford visit resulted in two books and several papers. One book [26] was jointly written with Aldo Andreotti who visited Stanford during the same year. This book is concerned with the dependence of meromorphic functions and has no connection with value distribution theory or the work of Chern. The other book [27] is inspired by the work of Chern as well as the results of H. Wu [29] and John Hirschfelder [17], [18]. Let M, N, F and G be connected, complex manifolds and $f : M \to N$ and $\pi : F \to N$ and $\rho : F \to G$ be holomorphic maps. We assume that N, F and G are compact and that G is a Kaehler manifold. We assume that π and ρ are surjective and have pure fiber dimensions u and p respectively. Then $\dim F = p + \dim G = u + \dim N$. For each $a \in G$, we assume that the map $\pi : \rho^{-1}(a) \to \pi\big(\rho^{-1}(a)\big) = E_a$ is biholomorphic. Take $z \in F$, we assume that there are open neighborhoods U of z and V of $\pi(z)$ and a reduced complex space W, such that there is biholomorphic map $\alpha : U \to V \times W$ and a holomorphic map $\beta : U \to W$ such that $\alpha(x) = \big(\alpha(x), \beta(x)\big)$ for all $x \in U$. If so, then $\{E_a\}_{a\in G}$ is called an *admissible target family* for f. Then

$$F_f = \{(x, y) \in M \times F | f(x) = \pi(y)\} \tag{75}$$

is an analytic subset of $M \times F$ and a complex manifold. The projections $\hat{f} : F_f \to F$ and $\pi_f : F_f \to M$ and the map $\rho_f = \rho \circ \hat{f}$ are holomorphic. The diagram (76) commutes:

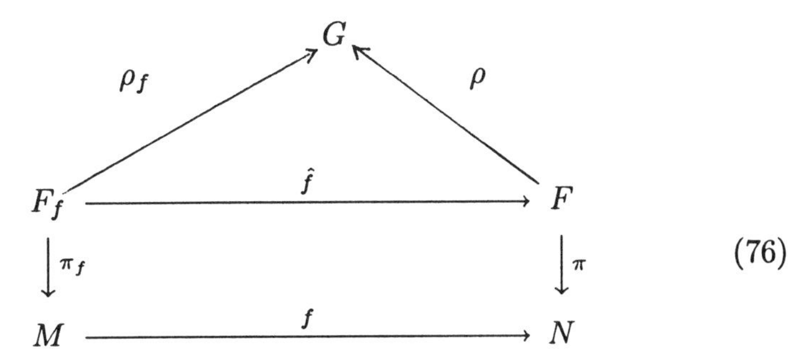

(76)

Let $\Theta > 0$ be the Kaehler volume form on G normalized that $\int_G \Theta = 1$. Then $\Omega = \pi_* \rho^*(\Theta) \geq 0$ is a form of bidegree (k,k) with $k = (\dim N) - p$. Here π_* is the fiber integration operator. According to Hirschfelder [17], [18], H. Wu [39] and Stoll [29], for every $a \in G$, there is a non-negative form λ_a of class C^∞ on $G - \{a\}$ such that $dd^c \lambda_a = \Theta$ on $G - \{a\}$. Then we have

$$\Lambda_a = \pi_*\big(\rho^*(\lambda_a)\big) \geq 0 \qquad \Omega = dd^c \Lambda_a \quad \text{on} \quad N - E_a \,. \tag{77}$$

The singularity of λ_a is constructed such that

$$\int_N \Omega \wedge \xi = \int_{E_a} \xi \qquad \text{for all} \quad a \in G \tag{78}$$

for any form ξ of bidegree (p,p) and class C^∞ with $d\xi = 0$ on N. Thus Ω is the Poincaré dual of the family $\{E_a\}_{a \in G}$. Basically we continue as with Diagram 8 and obtain a First Main Theorem and a Casorati-Weierstrass Theorem. Later Chia Chi Tung [37] extended this theory to the case where M, N, F, F_f are reduced complex spaces. The many technical difficulties in this generalization should not be underestimated. Tung's work is especially valuable, because it establishes many concepts and their properties on complex spaces (multiplicities, fiber integration, differential forms, Stokes Theorem). The standard properties of the fiber integration operator were known before I wrote [29], but proofs were difficult to find. Shiing-Shen Chern encouraged me to provide them in Appendix II of [29].

In 1965 Bott and Chern [3], [4] introduced vector bundles to value distribution theory, which proved to be most valuable. Let $\pi : W \to M$ be a holomorphic vector bundle of fiber dimension m over the connected, complex manifold M of dimension m. Let $0 : M \to W$ be the zero section. Put $W_* = W - 0(M)$. Let κ be a hermitian metric along the fibers of W. For $j = 0, 1, \ldots, m$ the Chern forms $c_j(W, \kappa)$ are defined. In a deep and complicated proof they construct a form ρ of bidegree $(m-1, m-1)$ and of class C^∞ on W_* such that $\pi^*\big(c_m(W,\kappa)\big) = dd^c\rho$ on W_* where ρ has the appropriate singularity on 0. Let $\mathfrak{a} : M \to W$ be a section of class C^∞ transversal to 0. For each $x \in M$ the intersection multiplicity $\nu_{\mathfrak{a}}(x)$ is defined. Then $\{x \in M | \nu_{\mathfrak{a}}(x) \neq 0\} = 0(M) \cap \mathfrak{a}(M)$ is a closed set of isolated points in M. Let $H \neq \emptyset$ be an open subset such that $\overline{H}$ is compact and such that $\partial H = \overline{H} - H$ is a smooth boundary manifold of H with $\mathfrak{a}(x) \neq 0(x)$ for all $x \in \partial H$. They establish the Unintegrated First Main Theorem

$$\int_H c_m(W,\kappa) = \sum_{x \in H} \nu_{\mathfrak{a}}(x) + \int_{\partial H} \mathfrak{a}^*(d^c\rho) \,. \tag{79}$$

If $H = M$ is compact, then $\partial H = \emptyset$ and the boundary integral disappears. Now assume, that M is not compact, but carries an exhaustion u of class C^∞ such that $dd^c u \leq 0$ outside a compact subset of M. Then u is called a *pseudoconcave exhaustion*. Also assume that $\mathfrak{a}$ is a holomorphic section of W over M such that supp $\mathfrak{a} = 0(M) \cap \mathfrak{a}(M)$ is a closed set of isolated points. Then $\nu_{\mathfrak{a}}(x) \geq 0$ is defined for all $x \in M$. For all $r > 0$, the *characteristic function* T, the *valence function* $N_{\mathfrak{a}}$ and the *deficit* $\Delta_{\mathfrak{a}}$ are defined by

$$T(r) = \int_0^r \int_{M[t]} c_m(W,\kappa)dt \qquad 0 \leq N_{\mathfrak{a}}(r) = \int_0^r \sum_{x \in M[t]} \nu_{\mathfrak{a}}(x)dt \tag{80}$$

$$\Delta_{\mathfrak{a}}(r) = \int_{M[r]} dd^c u \wedge \mathfrak{a}^*(\rho) \,. \tag{81}$$

For all $r \in \mathfrak{C}_u$, the *compensation function* $m_{\mathfrak{a}}$ is defined by

$$m_{\mathfrak{a}}(r) = \int_{M\langle r \rangle} d^c u \wedge \mathfrak{a}^*(\rho) \,. \tag{82}$$

For all $r \in \mathfrak{C}_u$, they establish the *First Main Theorem*

$$T(r) = N_{\mathfrak{a}}(r) + m_{\mathfrak{a}}(r) - \Delta_{\mathfrak{a}}(r) \tag{83}$$

which extends $m_{\mathfrak{a}}$ to a continuous function on $\mathbb{R}^+$ such that (83) holds for all $r > 0$. Let $\iota_r : M\langle r\rangle \to M$ be the inclusion map for each $r \in \mathfrak{C}_u$. Assume that the curvature form of (W, κ) is positive, then $c_m(W,\kappa) \geq 0$ and $\rho \geq 0$. Hence $\iota_r(d^c u \wedge \mathfrak{a}^*(\rho)) \geq 0$. Therefore $T(r) \geq 0$ and $m_{\mathfrak{a}}(r) \geq 0$. Also $dd^c u \wedge \mathfrak{a}^*(\rho) \leq 0$ outside a compact subset of M. Thus $\Delta_{\mathfrak{a}}$ is bounded from above and they obtain

$$N_{\mathfrak{a}}(r) \leq T(r) + 0(1) \ . \tag{84}$$

Of course $0(1)$ depends on $\mathfrak{a}$. For a Casorati-Weierstrass type theorem Bott and Chern need more assumptions: Let V be a finite dimensional vector space of global holomorphic sections of W. They assume that $\dim V = n + 1 > 1$ and that V spans W. A surjective bundle map $\eta : M \times V \to W$ is defined by $\eta(x, \mathfrak{a}) = \mathfrak{a}(x)$ for each $x \in M$ and $\mathfrak{a} \in V$. Also they require that there is at least one $\mathfrak{a}_0 \in V$ and $x_0 \in X$ such that x_0 is an isolated point of the zero set of $\mathfrak{a}_0$. Actually $N_{\mathfrak{a}}, m_{\mathfrak{a}}$ and $\Delta_{\mathfrak{a}}$ depend only on the projective value $a = \mathbb{P}(\mathfrak{a})$ of the section $\mathfrak{a}$. The First Main Theorem implies

$$0 \leq \delta(a) = 1 - \limsup_{r\to\infty} \frac{N_{\mathfrak{a}}(r)}{T(r)} \leq 1 \tag{85}$$

where $\delta(a)$ is called the *defect* of a. The *equidistribution theorem of Bott and Chern* states that $\delta(a) = 0$ for almost all $a \in \mathbb{P}(V)$.

The assumption that W is spanned by V allows us to simplify the proof considerably and reduce the proof to the case of an admissible family. Also the fiber dimension k of W can be different from the dimension of the base space and the zero set of a section can have positive dimension and may intersect the boundary. We can consider pseudoconvex manifolds M and take W over a connected complex manifold

N and consider the zero varieties as targets for a holomorphic map $f : M \to N$. This program was carried out in [30]. Moreover in [33] the theory was extended to all Schubert varieties of W.

Let me outline this extension of the Bott-Chern theory. Take integers p and n with $0 \leq p < n$. Put $k = n - p$. Let $\mathfrak{S}(p, n)$ be the set of all increasing maps $\alpha : \mathbb{Z}[0, p] \to \mathbb{Z}[0, k]$. Each element of $\mathfrak{S}(p, n)$ is called a *symbol*. For $\alpha \in \mathfrak{S}(p, n)$ define $|\alpha| = \alpha(0) + \cdots + \alpha(p) \in \mathbb{Z}_+$. Define $\hat{\alpha} : \mathbb{Z}[0, p] \to \mathbb{Z}[0, n]$ by $\hat{\alpha}(q) + q$ for all $q \in \mathbb{Z}[0, p]$. If $\alpha \in \mathfrak{S}(p, n)$ and $\beta \in \mathfrak{S}(p, n)$ define a partial ordering $\alpha \leq \beta$ by $\alpha(q) \leq \beta(q)$ for all $q \in \mathbb{Z}[0, p]$. For $\alpha \in \mathfrak{S}(p, n)$ define

$$G_\alpha(V) = \prod_{q=0}^{p} G_{\hat{\alpha}(q)}(V) \tag{86}$$

$$F(\alpha) = \{v = (v_0, \ldots, v_p) \in G_p(V) | E(v_0) \subset E(v_1) \subset \cdots \subset E(v_p)\} . \tag{87}$$

The *flag manifold* $F(\alpha)$ is compact and connected and has dimension

$$\dim F(\alpha) = \sum_{q=0}^{p} \big(n - \hat{\alpha}(q)\big)\big(\hat{\alpha}(q) - \hat{\alpha}(q-1)\big) = d(\alpha) . \tag{88}$$

The *Schubert family* $S(\alpha)$ of symbol $\alpha \in \mathfrak{S}(p, n)$ is the set of all points $(x, v) \in G_p(V) \times F(\alpha)$ with $v = (v_0, \ldots, v_p)$ such that

$$\dim E(x) \cap E(v_q) \geq q + 1 \qquad \text{for} \quad q = 0, 1, \ldots, p . \tag{89}$$

Then $S(\alpha)$ is an irreducible, analytic subset of dimension $d(\alpha) + |\alpha|$. The projections $\pi : S(\alpha) \to G_p(V)$ and $\sigma : S(\alpha) \to F(\alpha)$ are surjective. For each $x \in G_p(V)$ and $v \in F(\alpha)$, the sets

$$S(v, \alpha) = \pi\big(\sigma^{-1}(v)\big) \quad \text{and} \quad S_x(\alpha) = \sigma\big(\pi^{-1}(x)\big) \tag{90}$$

are irreducible analytic sets with $\dim S(v, \alpha) = |\alpha|$ and $\dim S_x(\alpha) = d(\alpha) + |\alpha| - d(p, n)$. The families $\{S(v, \alpha)\}_{v \in F(\alpha)}$ and $\{S_x(\alpha)\}_{x \in G_p(V)}$

are admissible. The analytic set $S(v,\alpha)$ is called the *Schubert Variety for the flag v and the symbol α.* Obviously

$$S(v,\alpha) = \{x \in G_p(V) | \dim E(x) \cap E(v_q) \geq q+1 \text{ for } q \in \mathbb{Z}[0,p]\} . \quad (91)$$

The *tautological bundle*

$$S_p(V) = \{(x,\mathfrak{z}) \in G_p(V) \times V | \mathfrak{z} \in E(x)\} \quad (92)$$

is a holomorphic subbundle of the trivial vector bundle $G_p(V) \times V$. The quotient bundle $Q_p(V)$ is defined by the exact sequence

$$0 \longrightarrow S_p(V) \underset{j}{\longrightarrow} G_p(V) \times V \underset{\eta}{\longrightarrow} Q_p(V) \longrightarrow 0 \quad (93)$$

where j is the inclusion and where η is the residual map. Here $S_p(V)$ and $Q_p(V)$ have fiber dimensions $p+1$ and k respectively. If $\alpha \in \mathfrak{S}(p,n)$ and $v = (v_0, \ldots, v_p) \in F(\alpha)$, then

$$S(v,\alpha) = \{x \in G_p(V) | \dim \eta_x\big(E(v_q)\big) \leq \alpha(q) \quad \text{for all } q \in \mathbb{Z}[0,p]\} . \quad (94)$$

Take a hermitian metric κ on V. It induces hermitian metrics along the fibers of $G_p(V) \times V$, $S_p(V)$ and $Q_p(V)$. These induced hermitian metrics are again denoted by κ. The qth Chern form of $Q_p(V)$ for κ is non-negative and is abbreviated by $c_q[p] = c_q\big(Q_p(V),\kappa\big) \geq 0$. Take $\alpha \in \mathfrak{S}(p,n)$. For $q \in \mathbb{Z}[0,p]$, let $\pi_q : F(\alpha) \to G_{\hat{\alpha}(q)}(V)$ be the projection. Cowen [13] introduced the positive volume form

$$\Omega_\alpha = \bigwedge_{q=0}^{p} \pi_q^*\big(c_{n-\hat{\alpha}(q)}[\hat{\alpha}(q)]^{\hat{\alpha}(q)-\hat{\alpha}(q-1)}\big) \quad (95)$$

of class C^∞ on $F(\alpha)$, where

$$\int_{F(\alpha)} \Omega_\alpha = 1 . \quad (96)$$

He defines the *Chern form* $c(\alpha) \geq 0$ for the symbol α by

$$c(\alpha) = \pi_* \sigma^*(\Omega_\alpha) \geq 0 . \quad (97)$$

Then $c(\alpha) \geq 0$ is the Poincaré dual of $S(v, \alpha)$ for all $v \in F(\alpha)$. The form $c(\alpha)$ has bidegree $\left(d(p,n) - \overrightarrow{\alpha}, d(p,n) - \overrightarrow{\alpha}\right)$. For $a \in \mathbb{Z}[0,p]$, and $b \in \mathbb{Z}[0,p]$, abbreviate $c_{ab} = c_{n-p-\alpha(a)-a+b}$ *Giambelli's theorem* [16] states

$$c(\alpha) = \begin{vmatrix} c_{oo}, & \ldots, & c_{op} \\ \vdots & & \vdots \\ c_{po}, & \ldots, & c_{pp} \end{vmatrix} . \tag{98}$$

The hermitian metric κ on V defines the unitary group $\mathfrak{U}(V)$ which acts transitively on $G_p(V)$ and $F(\alpha)$. Here Ω_α and $c(\alpha)$ are invariant under these group actions.

If $\alpha \in \mathfrak{S}(p,n)$, then $\alpha^* \in \mathfrak{S}(p,n)$ is defined by $\alpha^*(q) = n-p-\alpha(p-q)$ for all $q \in \mathbb{Z}[0,p]$. Obviously $|\alpha| + |\alpha^*| = (p+1)(n-p) = d(p,n)$. According to H. Wu [39], Hirschfelder [17], [18] and Stoll [29], for every $v \in F(\alpha)$ there exists a non-negative form λ_v of class C^∞ and bidegree $\left(d(\alpha)-1, d(\alpha)-1\right)$ on $F(\alpha) - \{v\}$ such that $dd^c\lambda_v = \Omega_\alpha$ on $F(\alpha) - \{v\}$ and such that

$$\int_{F(\alpha)} \lambda_v \wedge dd^c\psi + \psi(v) = \int_{F(\alpha)} \psi\Omega_\alpha \tag{99}$$

for all functions ψ of class C^∞. Moreover $g^*(\lambda_{gv}) = \lambda_v$ for all $g \in \mathfrak{U}(V)$. Then $\Lambda_v = \pi_*\sigma^*(\lambda_v)$ is non-negative form of class C^∞ and bidegree $(|\alpha^*| - 1, |\alpha^*| - 1)$ on $G_p(V) - S(v, \alpha)$ where $dd^c\Lambda_v = c(\alpha)$. If h is a vector function on $F(\alpha)$, define

$$\mathfrak{M}_\alpha(h) = \int_{F(\alpha)} h\Omega_\alpha . \tag{100}$$

Then $\hat{\lambda} = \mathfrak{M}_\alpha(\lambda_v)$ is a form bidegree $\left(d(\alpha)-1, d(\alpha)-1\right)$ on $F(\alpha)$ with $g^*(\hat{\lambda}) = \hat{\lambda}$ for all $g \in \mathfrak{U}(V)$. Hence $\hat{\lambda}$ is of class C^∞. Define

$$\Gamma(\alpha) = \{\beta \in \mathfrak{S}(p,n) | \beta \geq \alpha \quad \text{and} \quad \overrightarrow{\beta} = \overrightarrow{\alpha} + 1\} . \tag{101}$$

By a Theorem of Matsushima [22] (also see Stoll [32] and Damon [14], [15]) there are unique numbers $\gamma_{\alpha\beta} \geq 0$ for all $\beta \in \Gamma(\alpha)$ such that

$$\hat{\Lambda} = \pi_*\sigma^*(\hat{\lambda}) = \sum_{\beta \in \Gamma(\alpha)} \gamma_{\alpha\beta} c(\beta) \geq 0 . \tag{102}$$

Trivially $\hat{\Lambda} = \mathfrak{M}_\alpha(\Lambda_v)$. The proofs are total different. Matsushima uses Lie Algebras, and results of Kostant. I use fiber integration only. Damon uses algebraic topology, the Gysin homomorphism computed by a residue calculus.

Let M and N be connected, complex manifolds of dimensions m and $\tilde{n}$ respectively. Let W be a holomorphic vector bundle of fiber dimension $k \geq 1$ over N. Let $f : M \to N$ be a holomorphic map. Assume that there is given a complex vector space V of holomorphic sections of W with $\infty > \dim V = n + 1 > k$. Assume that V spans W. Put $p = n - k \geq 0$. Define $\eta : N \times V \to W$ by $\eta(x, \mathfrak{a}) = \mathfrak{a}(x)$ for all $x \in N$ and $\mathfrak{a} \in V$. Then η is a surjective bundle map and $S_W(V) = \ker \eta$ is a holomorphic vector bundle of fiber dimension $p+1$ over N. Let $j : S_W(V) \to N \times V$ be the inclusion map. We have the exact sequence

$$0 \longrightarrow S_W(V) \underset{j}{\longrightarrow} N \times V \underset{\eta}{\longrightarrow} W \longrightarrow 0 \,. \tag{103}$$

For $\alpha \in \mathfrak{S}(p, n)$ and $v = (v_0, \ldots, v_p) \in F(\alpha)$, the *Schubert Variety*

$$S_W(v, \alpha) = \bigcap_{q=0}^{p} \{x \in N | \dim \eta_*(E(v_q)) \leq a_q\} \tag{104}$$

of W is an analytic subset of N.

Take a hermitian metric κ on V which induces hermitian metrics – also denoted by κ- along the fibers of $N \times V$, $S_W(V)$ and W. Abbreviate $c_{ij} = c_{k-\alpha(i)-i+j}(W, \kappa)$ for all $i \in \mathbb{Z}[0, p]$, $j \in \mathbb{Z}[0, p]$. The *Chern form* $c_W(\alpha)$ of (W, κ) for $\alpha \in \mathfrak{S}(p, n)$ is defined by

$$c_W(\alpha) = \begin{vmatrix} c_{oo}, & \ldots, & c_{op} \\ \vdots & & \vdots \\ c_{po}, & \ldots, & c_{pp} \end{vmatrix} \,. \tag{105}$$

For each $x \in N$, the map $\eta_x : V \to W_x$ is linear and surjective. Hence $\ker \eta_x$ is a linear subspace of dimension $n + 1 - k = p + 1$.

Therefore one and only one $\varphi(x) \in G_p(V)$ exists such that $\ker \eta_x = E(\varphi(x))$. The map $\varphi : N \to G_p(V)$ is holomorphic and φ is called the classifying map for W. It pulls back the exact sequence (93) to the exact sequence (103) including the hermitian metric along the fibers. Thus we have

$$S_W(v, \alpha) = \varphi^{-1}(S(v, \alpha)) \qquad \text{for } v \in F(\alpha), \alpha \in \mathfrak{S}(p, n) \quad (106)$$

$$c_j(W, \kappa) = \varphi^*(c_j(Q_p(V), \kappa) \geq 0 \quad \text{for } j = 0, 1, \ldots, k \quad (107)$$

$$c_W(\alpha) = \varphi^*(c(\alpha)) \qquad \text{for } \alpha \in \mathfrak{S}(p, n) \, . \quad (108)$$

For each $v \in F(\alpha)$ with $\alpha \in \mathfrak{S}(p, n)$ a form

$$\Lambda_{W,v} = \varphi^*(\Lambda_v) \geq 0 \quad (109)$$

of class C^∞ and bidegree $(|\alpha^*| - 1, |\alpha^*| - 1)$ is defined on $N - S_W(v, \alpha)$ where we have $dd^c \Lambda_{W,v} = c_W(\alpha)$. Also

$$0 \leq \hat{\Lambda}_W = \mathfrak{M}_\alpha(\Lambda_{w,v}) = \varphi^*(\hat{\Lambda}) = \sum_{\beta \in \Gamma(\alpha)} \gamma_{\alpha\beta} c_W(\beta) \, . \quad (110)$$

Thus the value distribution theory for the map $f : M \to N$ in respect to the target family $\{S_W(v, \alpha)\}_{v \in F(\alpha)}$ coincides with the value distribution theory for the map $\varphi \circ f : M \to G_p(V)$ in respect to the admissible family $\{S(v, \alpha)\}_{v \in F(\alpha)}$.

We assume that $q = m + |\alpha| - k(p + 1) \geq 0$. We abbreviate

$$P_{v\alpha} = f^{-1}(S_W(v, \alpha)) \quad (111)$$

for $v \in F(\alpha)$ where $\alpha \in \mathfrak{S}(p, n)$ is fixed. We assume that $P_{v\alpha}$ is either empty or pure q-dimensional. (Actually it suffices that $P_{v\alpha} \neq \emptyset$ has dimension q at least one point $x \in P_{v\alpha}$ for at least one $v \in F(\alpha)$, but we make the stronger assumption for simplicity). For each $x \in P_{v\alpha}$ with $v \in F(\alpha)$ an intersection multiplicity $\nu_{f,v}(x) > 0$ is defined. If $x \notin P_{v\alpha}$, put $\nu_{f,v}(x) = 0$.

Let $u : M \to \mathbb{R}_+$ be a pseudoconvex exhaustion. Thus $dd^c u \geq 0$. We will assume that $dd^c u > 0$ on some non-empty open subset of M. The *spherical image function* $A_{\alpha,f}$ and the *characteristic function* $T_{\alpha,f}$ are defined for $r > 0$ by

$$A_{\alpha,f}(r) = \int_{M[r]} f^*(c(\alpha)) \wedge (dd^c u)^q \geq 0 \,, \; T_{\alpha,f}(r) = \int_0^r A_{\alpha,f}(t)dt \,. \tag{112}$$

The *counting function* and *the valence function* are defined by

$$n_{f,v}(r) = \int_{P_{v,\alpha}[r]} \nu_{f,v}(dd^c u)^q \geq 0 \qquad N_{f,v}(r) = \int_0^r n_{f,v}(t)dt \geq 0 \,. \tag{113}$$

For $r > 0$ the *deficit* is defined by

$$\Delta_{f,v}(r) = \int_{M[r]} f^*(\Lambda_{W,v}) \wedge (dd^c u)^{q+1} \geq 0 \,. \tag{114}$$

For $r \in \mathfrak{E}_u$, the *compensation function* $m_{f,v}$ is defined by

$$m_{f,v}(r) = \int_{M\langle r\rangle} f^*(\Lambda_{w,v}) \wedge d^c u \wedge (dd^c u)^q \geq 0 \,. \tag{115}$$

The theory of admissible target families yields the *First Main Theorem*

$$T_{\alpha,f}(r) = N_{f,v}(r) + m_{f,v}(r) - \Delta_{f,v}(r) \,, \tag{116}$$

which extends $m_{f,v}(r)$ to all $r > 0$ such that $m_{f,v}$ is continuous. Also

$$\begin{aligned} \mathfrak{M}_\alpha(m_{f,v}(r)) &= \int_{M\langle r\rangle} f^*(\Lambda_W) \wedge d^c u \wedge (dd^c u)^q \\ &= \int_{M[r]} f^*(\Lambda_W) \wedge (dd^c u)^{q+1} = \mathfrak{M}_\alpha(\Delta_{f,v}(r)) \,. \end{aligned} \tag{117}$$

Therefore we obtain

$$T_{\alpha,f}(r) = \mathfrak{M}_\alpha(N_{f,v}(r)) \,. \tag{118}$$

Define

$$B_f(\alpha) = \{v \in F(\alpha) | P_{v\alpha} \neq 0\} \qquad b_f(\alpha) = \int_{B_f(\alpha)} \Omega_\alpha \,. \tag{119}$$

Observe that $0 \leq b_f(\alpha) \leq 1$. Also

$$\begin{aligned}\int_{B_f(\alpha)} \Delta_{f,v}(r)\Omega_\alpha &\leq \mathfrak{M}_\alpha(\Delta_{f,v}(r)) \\ &= \int_{M[r]} f^*(\Lambda_W) \wedge (dd^c u)^{q+1} \\ &= \sum_{\beta\in\Gamma(\alpha)} \gamma_{\alpha\beta} \int_{M[r]} f^*(c_W(\beta)) \wedge (dd^c u)^{q+1} \\ &= \sum_{\beta\in\Gamma(\alpha)} \gamma_{\alpha\beta} A_{\beta,f}(r)\ .\end{aligned} \tag{120}$$

Also we have

$$\int_{B_f(\alpha)} N_{f,v}(r)\Omega_\alpha = \mathfrak{M}_\alpha(N_{f,v}(r)) = T_{\alpha,f}(r)\ . \tag{121}$$

Hence $\Delta_{f,v}(r) \geq N_{f,v}(r) - T_{\alpha,f}(r)$ implies

$$0 \leq 1 - b_f(\alpha) \leq \sum_{\beta\in\Gamma(\alpha)} \gamma_{\alpha\beta} \liminf_{r\to\infty} \frac{A_{\beta,f}(r)}{T_{\alpha,f}(r)}\ . \tag{122}$$

If $A_{\beta,f}(r)/T_{\alpha,f}(r) \to 0$ for $r \to \infty$ for all $\beta \in \Gamma(\alpha)$, then $b_f(\alpha) = 1$ which means that $f(M) \cap S_W(v,\alpha) \neq \emptyset$ for almost all $v \in F(\alpha)$.

The family $\{c(\alpha)\}_{\alpha\in\mathfrak{S}(p,n)}$ provides a set of free generators of the cohomology ring of the Grassmann manifold $G_p(V)$. Thus this extension of the theory of Bott and Chern includes all relevant invariants of the hermitian vector bundle W. The relation between the characteristic classes of Chern and the characteristic functions of Nevanlinna becomes clear. The name is a happy historically coincidence, which should not be destroyed by calling the characteristic function "height" all of a sudden. The work of Cowen [24] and the little blue book of Chern [29] later turned yellow, contributed much to this development.

For the Spring Semester 1973 Robert O. Kujala and Albert L. Vitter organized a program on Value Distribution in Several Complex Variables at Tulane University. The occasion was funded in part by a Science Development Grant from the National Science Foundation.

Michael Cowen and Ivan Cnop were there during the whole academic year. I participated during the second semester and gave a course on value distribution theory. Usual I was ahead with the research only two to four weeks before I lectured about it [20]. Invited visitors participated and lectured for about two weeks. In my experience this was my closest encounter with Wilhelm von Humbolds idea of an university. As I recall it, Shiing-Shen Chern was the first short term visitor. The Cherns had just returned from their first visit to China, after they had left the country in 1949. He talked about their visit and showed a great number of slides. It was very interesting. Subsequently the Cherns made many trips to China and he was a main mover to upgrade the mathematics in China. He founded a modern research institute near Beijing. I visited China only twice. In 1981 I participated in the Hangzhou Conference. In 1986 Marilyn and I visited for about 50 days under the auspices of the University of Science and Technology of China. I taught a course on Value distribution at Fudan and in Hefei. The students were transported from one place to another and I gave some talks at other institutions. Although a vice president at Fudan claimed that I teach too much, it was absolutely insignificant to Marilyn's teaching schedule at the Anhui Fine Arts Institute in Hefei, where she was the first foreigner to teach. The Institute was most grateful for her marathon performance in teaching the piano. These contacts gave us a closer look at Chern's home land, but we never met him in China.

The visit to Tulane was a wonderful experience and the conference was a great success. The results are documented in [30]. Chern, Cowen and Vitter [11] proved, that a Kaehler manifold permits a Frenet frame for each of its holomorphic curves if and only if the holomorphic sectional curvature is constant.

Frenet frames are most helpful to handle associated maps. For instance the Ta-operator interconnects the associated maps of the two meromorphic maps $f : M \to \mathbb{P}(V)$ and $g : M \to \mathbb{P}(V^*)$. Ten years after the Tulane conference, Frenet formulas helped me to prove the Stress Curvature Formula and the Stress Gradient Formula for the Ta-operator, [35] Theorems 5.9 and 10. The calculations for the iterated Ta-operator are very much more complicated and lead to a new type of defect relations. The results are published in [36], but since 1985 I did not find the time to prepare the manuscript for the proofs. Perhaps this has to wait until retirement. I hope that with the theories outlined above and with the two stress formulas I have complied – at least partially – with Chern's admonishment in Philadelphia to infuse differential geometry into value distribution.

After the Tulane conference, we met on several occasions, which I do not all recall. Three occasions stand out and are unforgettable. A huge crowd assembled for Chern's retirement conference in Berkeley 1979. The talks were excellent and there was a remarkable dinner session in Chern's favorate Chinese restaurant. About 10 years later, I was able to persuade him to participate in a small Symposium on Value Distribution Theory in Several Complex Variables at Notre Dame, April 28–29, 1990. He attended all the lectures and offered many pertinent remarks, but he declined to lecture, because he felt that he is not active in this area anymore. Mrs. Chern came with him. By coincidence, the Physics department had invited Paul Chu, Chern's son-in-law, to deliver a lecture on superconductivity the day before the symposium. Chern attended this lecture too and they had a short family reunion. At the Symposium dinner, Shiing-Shen Chern told me the most demanding task in the organisation of the symposium was still before me: the production of the proceedings. He was correct! After the Symposium the Cherns proceeded on their travel to Europe, where he was inducted into the French Academy of Science, an honor well deserved. A few weeks later I met him at a conference in Los Angeles on the occasion of his 80th birthday.

Recently, I congratulated my brother Hans to his 65th birthday.

In my letter, I reflected upon my own experience of this special day and remarked: "Man kann selbst garnicht begreifen, dass man schon so alt ist." I wish that the Cherns have and will retain the same feeling for a long time to come and that they will spend many happy years together.

References

1. Ahlfors, L., *The theory of meromorphic curves.*, Acta. Soc. Sci. Fenn. Nova Ser. **A3(4)** (1941), 171–183.
2. Andreotti, A. and W. Stoll, *Analytic and algebraic dependence of meromorphic functions*, Lecture Notes in Math **234** (1971), p.p.390, Springer-Verlag.
3. Bott, R. and S.S. Chern, *Vector bundles and the equidistribution of the zeroes of their holomorphic sections.*, Acta Math **114** (1965), 71–112.
4. Bott, R. and S.S. Chern, *Some formulas related to complex transgression*, Essays on Topology and Related Topics (1970), 48–57, Springer-Verlag.
5. Chern, S.S., *The integrated form of the first main theorem for complex analytic mappings in several complex variables.*, Annals of Math **71** (1960), 536–551.
6. Chern, S.S., *Complex analytique mappings of Riemann surfaces I.*, Amer. J. Math. **82** (1960), 323–337.
7. Chern, S.S., *Holomorphic mappings of complex manifolds.*, L'Ens. Math **7** (1961), 179–187.
8. Chern, S.S., *Complex manifolds without potential theory. Van Nostrand*, (1968) p.p. 62 Second Edition; Universitext (1979), p.p.62, Springer-Verlag.
9. Chern, S.S., *On holomorphic mappings of hermitian manifolds of the same dimension.*, Proc. of Symp. in Pure Math. **11** (1968), 157–170.
10. Chern, S.S., *Holomorphic curves and minimal surfaces*, Carolina Conference Proceedings (1970), p.p. 28.
11. Chern, S.S., *Holomorphic curves in the plane*, Diff. Geom. in Honor of K. Yano. Kinokuniya, Tokyo (1972), 73–94.
12. Chern, S.S., Cowen, M. and Al Vitter, *Frenet frames along holomorphic curves*, Proc. of Conf. on Value Distribution Theory. Tu-

lane University (1974), 191–203 (see [20]A)..

13. Cowen, M., *Hermitian vector bundles and value distribution for Schubert cycles.*, Trans. Amer. Math. Soc. **180** (1973), 189–228.
14. Damon, J., *The Gysin homomorphism for flag bundles.*, Amer. J. of Math. **95** (1973), 643–659.
15. Damon, J., *The Gysin homomorphism for flag bundles.*, Applications. Amer J. of Math **96** (1974), 248–260.
16. Gambelli, G.Z., *Risoluzione del problema degli spazi secanti*, Mem. R. Acc. Torino **52** (1902), 171–211.
17. Hirschfelder, J., *The first main theorem of value distribution in several variables*, Invent. Math **8** (1969), 1–33.
18. Hirschfelder, J., *On Wu's form of the first main theorem of value distribution*, Proc. Amer. Math. Soc. **23** (1969), 548–554.
19. Kneser, H., *Zur Theorie der gebrochenen Funktionen mehrerer Veränderlichen*, Jber. Deutsch Math Verein. **48** (1938), 1–38.
20. Kujala, R. and A. Vitter III, *Value-Distribution Theory*, Marcel Dekker Part A, Pure and Appl. Math **25** (1974), p.p.269, Part B: W. Stoll: Deficit and Bezout Estimates (1973), p.p.271.
21. Levine, H., *A theorem on holomorphic mappings into complex projective space.*, Ann. of Math(2) **71** (1960), 529–535.
22. Matsushima, Y., *On a problem of Stoll concerning a cohomology map from a flag manifold into a Grassmann manifold.*, Osaka J. Math **13** (1976), 231–269.
23. Molzon, R.E., B. Shiffman and N. Sibony, *Average growth estimates for hyperplane sections of entire analytic sets.*, Math. Ann **257** (1981), 43–59.
24. Stoll, W., *Die beiden Hauptsätze der Wertverteilungstheorie bei Funktionen mehrerer komplexen Veränderlichen*, Acta Math **I 90** (1953), 1–115, **II 92** (1954), 55–164.
25. Stoll, W., *The multiplicity of a holomorphic map*, Invent. Math. **2** (1966), 15–58.

26. Stoll, W., *The continuity of the fiber integral.*, Math Zeitschr **95** (1967), 87–138.
27. Stoll, W., *A general first main theorem of value distribution*, Acta Math **118** (1967), 111–191.
28. Stoll, W., *About the value distribution of holomorphic maps into projective space.*, Acta Math. **123** (1969), 83–114.
29. Stoll, W., *Value distribution of holomorphic maps into compact complex manifolds.*, Lecture Notes in Math **135** (1970), p.p.267, Springer-Verlag.
30. Stoll, W., *Value distribution of holomorphic maps.*, Several Complex Variables I Maryland 1970, Lecture Notes in Math **155** (1970), 165–190, Springer-Verlag.
31. Stoll, W., *Value distribution on parabolic spaces*, Lecture Notes in Math. **600** (1977), p.p.216, Springer-Verlag.
32. Stoll, W., *Invariant forms on Grassmann manifolds*, Ann. of Math. Studies **89** (1977), p.p.113, Princeton University Press, Princeton, N.J..
33. Stoll, W., *A Casorati-Weierstrass theorem for Schubert zeros of semi-ample holomorphic vector bundles.*, Atti Acad. Naz. Lincie. Mem. C1 Sci. Fis. Mat. Natur. Ser VIIIm **15** (1978), 63–90.
34. Stoll, W., *The characterization of strictly parabolic manifolds.*, Ann Scuola. Norm. Pisa **7** (1980), 87–154.
35. Stoll, W., *Value distribution theory for meromorphic maps.*, Aspect of Math. **E7** (1985), p.p.347, Vieweg-Verlag.
36. Stoll, W., *Value distribution theory for moving targets. Complex Analysis and algebraic geometry. Proceedings*, Lecture Notes in Math **1194** (1986), 214–235, Göttingen.
37. Tung, C.C., *The first main theorem of value distribution on complex spaces.*, Atti della Acc. Naz d. Lincie Serie VIII **15** (1979), 93–262, Notre Dame Thesis 1973.
38. Weyl, H. and J. Weyl, *Meromorphic functions and analytic curves*, Annals of Math. Studies **12** (1943), p.p.269, Princeton University

Press, Princeton, N.J..

39. Wu, H., *Remarks on the first main theorem in equidistribution theory, I, II, III, IV*, J. Diff. Geometry **2** (1968), 197–202, 369–384, ibid. vol. **3** (1969), 83–94, 433–446.

40. Wu, H., *The equidistribution theory of holomorphic curves.*, Annals of Math Studies **64** (1970), p.p. 219, Princeton University Press, Princeton, N.J..

My Encounters with S.S. Chern

Wilhelm Klingenberg
Mathematisches Institut
Universität Bonn

More than four years I had to spend in the Wehrmacht during the Great War before I finally could begin to study Mathematics. I was then 21 years old, the universities in Göttingen and Hamburg were full in the Fall 1945, so I had to begin at Kiel University where K.H. Weise taught me the tricks of tensor calculus. I wrote my thesis on Affine Differential Geometry and continued with solving a problem of Schouten, Hlavaty, Laptev and others of rigging submanifolds in manifolds with affine connection. It was pretty esoteric stuff and only a few specialists took notice. However, thirty years later it helped me to get my first invitation to the Peoples Republic of China, where Su Buchin, from Fudan University in Shanghai, had taken a liking of my work.

There was no future for me in Kiel, but Weise helped me to get in contact with Wilhelm Blaschke in nearby Hamburg. Blaschke had been most influential to the development of Differential Geometry, and in the Twenties and Thirties Hamburg had been a Mekka for students all over the world who wanted to catch up with modern trends in Mathematics; witness the many papers in the "Hamburger Abhandlungen". Only Elie

Cartan in Paris could be compared in his influence with Blaschke. As the Third Reich went on, all this began rapidly to fall apart and after the war Blaschke never recovered for Hamburg the position it once held.

During the heydays, among the bright young students visiting Hamburg was also S.S. Chern. At Peking University he had met E. Sperner as visiting lecturer and this led him to go to Hamburg, where he wrote a thesis under Blaschke. After that, he went on to study in Paris under Cartan. Already in Hamburg he had learnt from E. Kähler the calculus of forms introduced by Cartan and during all his life he preferred the method of moving frames to the tensor calculus employed by the German, Russian and Dutch school. At that time some people tended to pay much attention to the methods, thereby neglecting an evaluation of the results. In my generation this had stopped to be an article of faith since everyone agreed that both methods were just dual aspects of the same object.

As a person educated in the refined traditions of his country, S.S. Chern always regarded his teachers with the greatest respect. Thus, he came back to Hamburg in 1953 to pay a visit to Blaschke. At that time, I had a position as an assistant in Applied Mathematics under L. Collatz – not a very fortunate situation, but nothing else could be found for me. I was there under a 2-year contract which was not continued. Blaschke then helped me to find a job in Göttingen under K. Reidemeister, which worked out very fine.

During his visit in Hamburg in the summer of 1953 I was introduced to Chern. My work so far cannot have made a great impression on Chern, I had also done something on Laguerre Geometry but mainly I had switched to Foundations of Geometry and Geometric Algebra. Still, the famous visitor treated me most kindly, Blaschke must have put in a few kind words for me. I still remember that Chern invited me to the coffee room of the most elegant hotel of Hamburg, the famous

"Vier Jahreszeiten". We also must have discussed mathematics – Chern is an inveterate talker always ready to discuss mathematical problems, even if he is not actively working himself on them. He has a deep insight into incredibly many different subjects. This justly brought him into the prominent position for the development of Modern Mathematics in our century. I could only admiringly listen to him while we were enjoying our delicacies in the elegant surroundings which I never had visited before and never since, to tell the truth.

I wonder what he must have thought of me then, but it seems he did not form an adverse opinion. Although I never became something of a student of him, he always treated me with kindness and I have no doubt that at more than one occasion he has helped my career behind the scenes.

Over the vicissitudes of the times under Hitler, Blaschke managed to save some of his old friendships, among them with M. Morse at the Institute for Advanced Study in Princeton. My application to become a visitor there was supported by Blaschke and doubtless by some other referees, maybe Chern among them. In any case, my two years in Princeton from 1956 to 1958 were the most influential for mathematical career. There I found some basic information in the cut locus of Riemannian Manifolds, the proof of the sphere theorem is based on them, among others.

My whole approach was so new and unusual, that neither Morse nor H. Rauch, who had worked on the problem before, believed it to be true. But Chern must soon have been convinced otherwise. Backfeeding in those years still took some time. While holding in Göttingen the position of an assistant, although with Habilitation which I had taken still in Hamburg, I was invited as a visiting full professor to Berkeley. I went there from January 1962 to January 1963, and here it was for the first time that I had prolonged continuous contact with Chern. My wife Christine still remembers the personal assistance he gave us when finding a house for us and our three small children – it was the house of H. Helson in Shattuck Avenue. In every respect, Chern treated me as a colleague, he never seemed to mind that I had not chosen for my work problems close to his own line of research. Back in Germany I became Professor in Mainz and then – in 1966 – in Bonn, after having declined an offer from the University of Zürich.

I continued to meet Chern. One of our common activities was the organization of meetings on Global Differential Geometry in Oberwolfach. These meetings which take place every two years have become a tradition which hopefully will continue also after we laid them in to the hands of younger colleagues.

My friendship with Chern reached a new dimension with the opening of China for foreign visitors. I had become fascinated with China and its history years ago; during my various trips to Japan I always had looked there for traces of Chinese civilization. As I mentioned already, my first visit to China took place in 1980. In 1984 I went there again, this time I had a chance to visit Dunhuang. Two years later began quite a series of visits to China, starting with a back packing tour along the legendary Silk Road. For and educated Chinese like Chern this must have seemed a little strange, but I had read in my boyhood the stories of the explorers and, after all, did I not follow the footsteps of the famous pilgrim Xuan Zhang in the time of the Tang dynasty? Already in 1980 I had visited his burial place south of Xian.

In 1988 I came to Nankai University and was an honoured guest in Chern's elegant house. From there I started for my first visit to Lhasa and beyond. The following year Tibet was closed to foreigners, but I could visit Sichuan province where I climbed the Omei shan and made extensive trips in to the mountainous regions further east up to the border with Tibet. Finally, 1990 I could make a pilgrimage to the Kangrinboqe Feng or Kailas in the far West of Tibet, a holy mountain for Buddhists and Hindus alike. Each time I came in close contact with the Chinese with whom I made many new friends.

Coming back to the main topic of my story, I can state that S.S. Chern was the first Chinese I met in my life. Although we never worked together in a paper, I feel that he has extended to me always generous friendship or, to say it more colloquial, he was my benevolent Godfather. In him I find embodied the true citizen of the world, a very rare species which can only thrive when it has its roots in the great traditions of wisdom and tolerance peculiar to his race.

On the normal Gauss map of a tight smooth surface in $\mathbb{R}^3$

Francois Haab
Departamento de Matematica
UFMG, 31270 Belo-Horizonte

Nicolaas H. Kuiper
Institut des Hautes Etudes Scientifiques
91440 Bures-sur-Yvette, and
Math. Inst. Universiteit, Utrecht.

To S.S. Chern in friendship and esteem.

The normal Gauss map of a tight smooth surface in $\mathbb{R}^3$ is reminiscent of minimal surfaces and holomorphic functions.

The outside unit normal vectors of a convex smooth (of class C^∞) surface $M \subset \mathbb{R}^3$ determine the Gauss map $g : M \to S^2$ into the unit sphere. For a strictly convex surface, respectively an analytic convex surface, g is a diffeomorphism, resp. a homeomorphism. But in general it is a *cell-like map*, namely, for which the inverse image $g^{-1}(z)$ for $z \in S^2$ is *cellular* because convex. It is the intersection of a nested sequence of open discs in M. We will generalize this to all tight surfaces in theorems 1,2,3. A more detailed proof is found in [H$_2$]. Analogous theorems hold for *locally tight surfaces.*

Let $f : M \to \mathbb{R}^3$ be a smooth immersion of a closed possibly non orientable surface, with *Euler number* χ. The total absolute curvature functional $\tau(f)$ obeys

$$\tau(f) = \frac{1}{2\pi} \int |\kappa \; d\,\sigma| \geq 4 - \chi \,. \tag{1}$$

The immersion is called *tight* if the infimum $\tau(f) = 4 - \chi$ is attained. For many years it was unknown whether the projective plane with one handle ($\chi = -1$) admits a smooth tight immersion, but recently Haab [H$_1$] proved it does not. For topological or polyhedral immersions (see definition B) the problem remains open.

In this note we present structure theorems, that simplify to some extend the subtle proof in [H$_1$]. We also clarify how complicated the Gauss map of a tight smooth surface can be locally. The *Gauss map* $g : M \to P = (S^2 \;/\; \text{antipodal})$ assigns to $x \in M$ the pair of antipodal unit normal vectors to the surface M at $f(x)$. These unit vectors form the unit sphere S^2 and the pairs form the projective plane P with non euclidean metric with curvature $\kappa = 1$. We refer often to a point of P by its "coordinate" $z \in S^2$.

If the linear function (or covector) z^* has gradient $z \in S^2$, then the composition $h_z = z^* \circ f$ is called a *height function* on M. The point $z = g(x)$ is Gauss map value of $x \in M$, if and only if h_z has $x \in M$ as a critical point. The set of values z for which h_z is a non degenerate function on M is open and dense in S^2 (or P) for any smooth immersion f.

Other equivalent and useful definitions of tightness for surfaces are given by the following properties. See [CR] and [K], also for generalizations.

Definition A. *Every non degenerate height function h_z has exactly $4 - \chi$ critical points.*

Definition B. *Every half space H (bounded by a plane in $\mathbb{R}^3$) meets M in a connected set $f^{-1}(H)$.*

Another interesting property is:

Property C. *The Gauss-curvature κ of a tight surface is $\kappa \geq 0$ on the boundary of the convex hull $\mathcal{H}\, f(M)$ of $f(M)$, and $\kappa \leq 0$ elsewhere on M.*

By definition A we see that the Gauss map of a tight surface is something like a $(4-\chi)$-fold covering of P. More precisely we have for tight f the

Theorem 1. *Outside a finite set $Z = \{z_1, \dots, z_m\} \subset P$ of values, the Gauss map g of a tight surface is a composition of a cell-like map γ and a$(4-\chi)$-fold covering β:*

$$M \setminus g^{-1}(Z) \underset{\gamma}{\rightarrow} N \underset{\beta}{\rightarrow} P \setminus Z \subset P\ .$$

We call the composition $\beta\gamma$ a *cell-like covering* of $P \setminus Z$. Clearly $M \setminus g^{-1}(Z)$ has one (convex) component where $\kappa \geq 0$ and g is a double cell-like covering onto $P \setminus Z$ on that convex component. Recall that if $\gamma : A \to B$ is a cell-like map between surfaces, then A and B are homeomorphic. A component K_i of the set K of critical points of a height function on a tight surface may have interior points. For example near a point where $\kappa \geq 0$. resp. $\kappa \leq 0$, a tight surface can be flattened while preserving tightness. For $\kappa > 0$ we can obtain a plane convex critical component. For $\kappa \leq 0$ we can obtain a flat quadrangle with four vertices near which the height above the tangent plane changes sign, and with concave sides. If not convex, resp. not concave, we easily contradict definition B of tightness. See figure 1 a) where we start from a tight standard round torus, placed vertical, and flattened at A and B to get a fat maximum and a fat saddle point in horizontal planes in figure 1 b). In figure 1 c) we attach a handle ending in these flat horizontal parts and obtain from the torus a tight smooth non orientable surface with $\chi = -2$. Figure 1 c') is a variant that illustrates theorem 2.

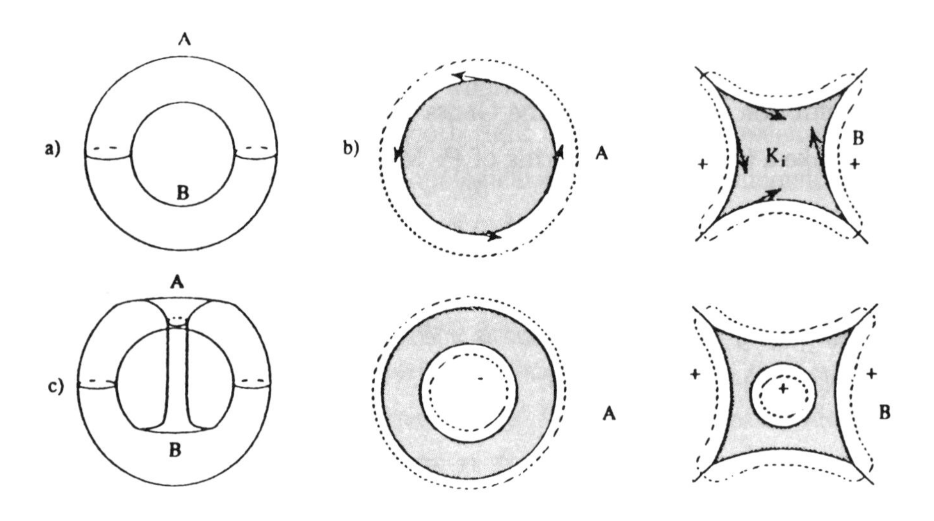

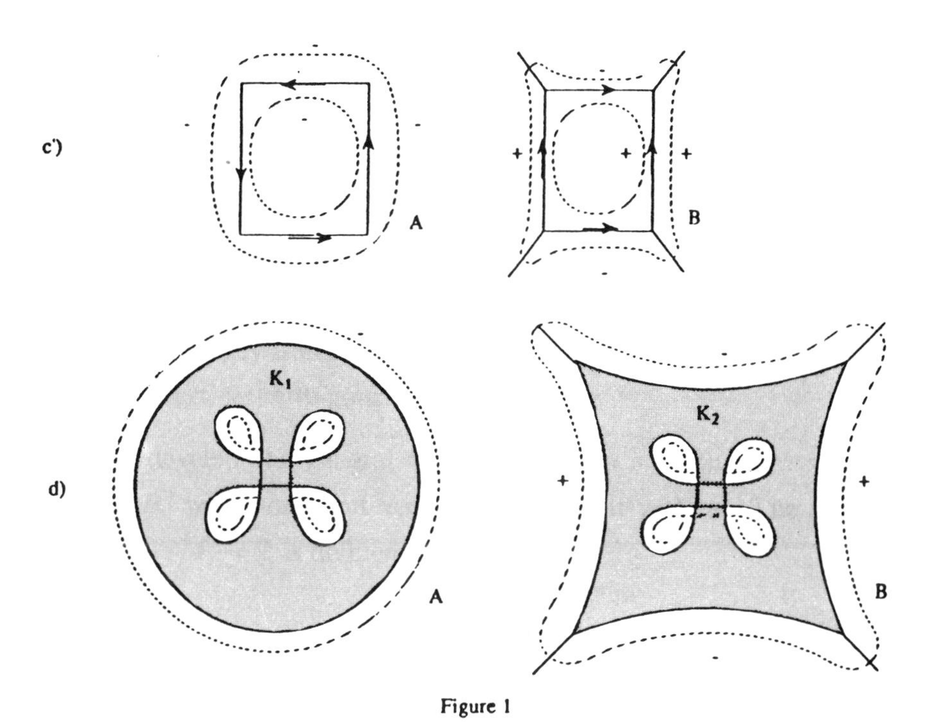

Figure 1

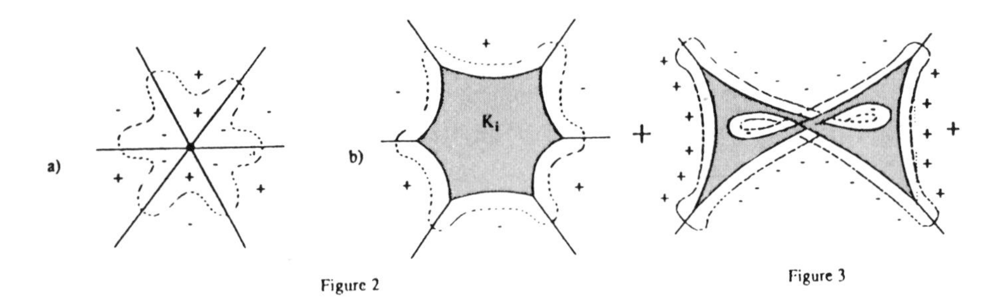

Figure 2

Figure 3

In figure 1 d) the two horizontal exceptional critical components are each a topologically immersed torus from which two open discs are deleted. These two critical horizontal components are further connected by a smooth immersed handle ($\kappa \leq 0$) to obtain a smooth tight non orientable surface with $\chi = -6$. In the figures we only show the critical (horizontal) components at A and B as seen from above.

In figure 2 a) we show a Monkey saddle point on a part of surface where $\kappa \leq 0$ and in figure 2 b) a fat Monkey saddle. In figure 3) we show a curious example of a critical height component K_i on a tight surface. It is a topologically immersed annulus bounded by two locally concave (except at vertices) two-gons.

We next develop the general theroy. Given a smooth tight surface $f : M \to \mathbb{R}^3$ we fix one unit vector z_0 and call it vertical. The *smooth function* $\alpha : M \to \mathbb{R}$ is defined by

$$\alpha(x) = \text{dist}(z_0, g(x)) \leq \pi/2 \text{ , dist=distance in } P \text{ .} \tag{2}$$

Let $K_1, \ldots , K_s$ be some finite set of disjoint (critical) components of the subspace K of critical points of the height function h_{z_0}.

The open round disc with radius $\varepsilon < \pi/2$ in P and with center z_0 is denoted U_ε, its boundary is ∂U_ε. The inverse image under g is $g^{-1}(U_\varepsilon) = U_\varepsilon(K)$ and $U_\varepsilon(K_i)$ is the component of $U_\varepsilon(K)$ that contains K_i. If $\varepsilon > 0$ is a *regular value* of the function α, then $\partial U_\varepsilon(K_i)$ is a union of imbedded circles in M, and $U_\varepsilon(K_i)$ is a surface in M with that union as boundary. We can assume $\varepsilon > 0$ small and so chosen that some value $z \in \partial U_\varepsilon$ has a *non degenerate* height function h_z on M. Near z the Gauss map is union of $4-\chi$ immersions of disjoint discs in M. The $\partial U_\varepsilon(K_i)$ produces for small $\varepsilon > 0$ at least one contribution ≥ 1 for $i = 1, \ldots , s$, to the number of critical points $(4 - \chi)$ of h_z. Therefore the number of components of K is at most $4 - \chi$ and in particular it is finite.

Let this number now be $s \leq 4 - \chi$. Then

$$K = \bigcup_{i=1}^{s} K_i \; . \tag{3}$$

By tightness we can conclude that g induces for small ε a *monotone map* on each of the circles of $\partial U_\varepsilon(K_i)$ into the circle $\partial U_\varepsilon \subset P$. We next examine M near $K_i \subset K$, where K_i is immersed in a horizontal plane.

Roberts and Steenrod [RS] proved for a continuum in a closed surface, like $K_i \subset M$, that $M \setminus K_i$ intersects some neighborhood U_i of $K_i \subset M$ in a disjoint union of a finite number, say j_i open annuli H_{ij}, $j = 1, \dots, j_i$, each bounded by a circle in M on one end, and bounded by a subset of K_i on the other open end of the annulus H_{ij}. As disjoint then so are $U_\varepsilon(K_i) \setminus K_i \subset U_j H_{ij}$ for small ε. We also (but first) chose H_{ij} so small that h_{z_0} has no critical point on H_{ij}. We then find in our context that for small ε the circles of $\partial U_\varepsilon(K_i)$ are embedded in such annuli H_{ij}. Note that for very small ε none of these circles bounds a disc in its annulus, because that disc would contain by virtue of the Gauss map another critical point of h_{z_0} not in K. So any two circles so obtained in H_{ij} are homotopic in H_{ij} and as $U_\varepsilon(K_i)$ is connected there is only one such circle, for given ε. It contributes a *multiplicity* $\mu_{ij} \geq 1$ equal to the degree of the covering by g onto ∂U_ε. Following the Gauss map image along a circle, it will go around z_0 in P μ_{ij} times. See the examples. With $\mu_i = \sum_{j=1}^{j_i} \mu_{ij}$, we have

$$\sum_{i=1}^{s} \mu_i = \sum_{i=1}^{s} \sum_{j=1}^{j_i} \mu_{ij} = 4 - \chi \; . \tag{4}$$

In figures 1 b), 1 c), 1 d), 2), 3) we have $\mu_i = 1, 1+1, 1+3, 2, 1+1$ respectively. Some circles in $\partial U_\varepsilon(K_i)$ are shown in orthogonal projection in the tangent plane at K_i. For small $\varepsilon > 0$ the orthogonal projection gives always an immersion of $U_\varepsilon(K_i)$ in that tangent plane. By taking a monotone sequence of regular values ε_u converging for $u \to \infty$ to zero we see that K_i is the intersection of a nested sequence $U_{\varepsilon_u}(K_i)$. The interior of the difference of two such consecutive surfaces has in H_{ij} an open annulus denoted $A_{ij} \subset H_{ij} \subset M$ with smooth boundaries A_{+ij} and A_{-ij}. It is called *thin* if $\varepsilon_u - \varepsilon_{u+1} > 0$ is small. In the limit the circles in H_{ij} converge to a convex curve or to a locally concave curve, possibly with vertices at which the height function above the tangent plane of K_i changes sign, possibly degenerated into a ciritcal set without interior points of dimension ≤ 1 as in figure 1 c', and possibly degenerated into one point as seen in figure 2 a). A critical component K_i of some height function is called *Morse-nonsingular* in case its multiplicity $\mu_i = \sum_{j=1}^{j_i} \mu_{ij}$ is *one*. Otherwise it is called *Morse-singular.*

Lemma 1. *If K_i is Morse-nonsingular then K_i is a cellular set.*

For a complete proof see [H_2]. Here we treat special cases and give a suggestive motivation for the validity of the lemma.

Let K_i have interior points and be a surface with as boundary one circle ∂K_i. If ∂K_i immerses by f onto a smooth locally convex curve $f\partial K_i$, then K_i is a disc with convex image. Hence K_i is cellular. There is on the other hand no surface K_i immersed in a plane for which $f\partial K_i$ is one smooth locally concave curve.

Now let $f\partial K_i$ be a piecewise smooth locally concave curve with 4 *cusps* at vertices ("cusp" means that the solid angle at the vertex is $\rho = 0$). See for example figure 4. Suppose K_i is the surface of genus γ_i, from which an open disc has been deleted. By considering the Gauss map g restricted to $\partial U_\varepsilon(K_i)$ for $\varepsilon > 0$ converging to zero, we conclude that μ_i is the rotation number of a continuous monotone one parameter family of (normal or) tangent unit vectors along $f\partial K_i$. The family is continuous also at the cusps. In figure 4, $\gamma_i = 1$ and $\mu_i = 2\gamma_i + 1 = 3$. Only the disc (for $\gamma_i = 0$) with image a fat saddle point gives a solution $\mu_i = 1$. See figure 1b B.

If we replace the smooth sides of $f(\partial K_i)$ by general concave sides there is no essential influence on the argument. If we replace (increase) a cusp angle $\rho = 0$ by a solid angle $\rho > 0$, $0 < \rho < 2\pi$, while preserving the local concavity on the sides of ∂K_i, then we must add (incorporate) the solid angles ρ to (into) the rotation number and μ_i does not change! With such isotopies we cannot easily cover all degenerate K_i's however. For a different approach and a complete proof see therefore [H₂].

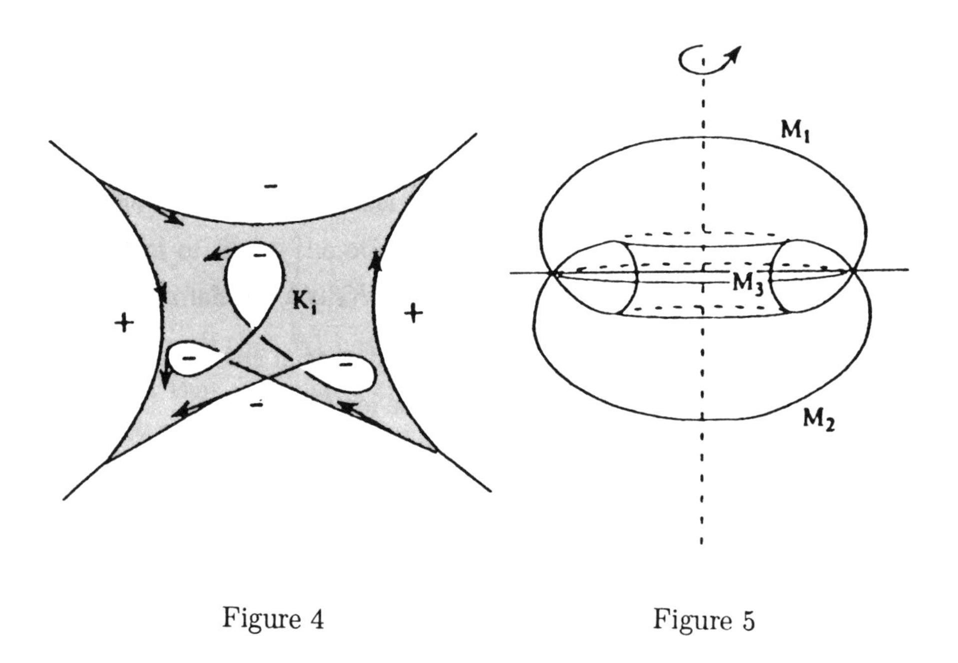

Figure 4

Figure 5

Lemma 2. *Morse-singular critical components of height functions are isolated in M.*

Corollary. *There are only a finite number of Morse-singular critical components of height functions in M.*

Proof of lemma 2. Let x be any point in $U_{\varepsilon_0}(K) \setminus K$ for small ε_0 and say in H_{ij}, and in a thin annulus A_{ij}:

$$x \in A_{ij} \subset H_{ij} \subset U_{\varepsilon_0}(K) \ .$$

Call $g(x) = z_1 \in P$, and study h_{z_1} like before h_{z_0}. For small suitable ε. and a very thin annulus A_{ij}, the round disc $U_\varepsilon(z_1) = \{z \in P : \text{dist}\ (z, z_1) < \varepsilon\}$ meets the round annulus $g(A_{ij})$ in a quadrangle $Q \subset P$. For ε a regular value of $y \to \text{dist}\ (g(y), z_1)$, we observe that $g^{-1}(\partial U_\varepsilon(z_1))$ meets A_{ij} in $2\mu_{ij}$ smooth curves that connect the boundary circls $\partial_+ A_{ij}$ and $\partial_- A_{ij}$ of A_{ij}. This implies that $g^{-1}U_\varepsilon(z_1) \cap A_{ij}$ consists of μ_{ij} quadrangles, each of which has a boundary that maps under a monotone map into ∂Q, of some degreee ≥ 1, hence of degree $=1$ as the degrees add up to $= \mu_{ij}$. The Gauss map g restricted to each of the μ_{ij} quadrangles is then cell-like, and the same holds for the μ_{ij} remaining quadrangles on A_{ij}, which are each mapped onto the remaining quadrangle of $g(A_{ij})$. Let $A_{ij} \to A_{ij}$ / sing be the quotient space of the identification relation sing which sends each critical cellular component A_{ij} into one point. Then $g|A_{ij}$ is a *cell-like covering of degree* μ_{ij}. It is a composition of a cell-like map γ and a covering map β of degree (=mulitplicity) μ_{ij}:

$$A_{ij} \underset{\gamma}{\to} A_{ij}/\text{sing} \underset{\beta}{\to} g(A_{ij}) \subset P\ .$$

For fixed i, the annuli H_{ij} can be covered by such annuli A_{ij}, which agree on their intersections and we find for ε small a composition of a cell-like map γ and a covering β of degree (multiplicity) μ_i :

$$U_\varepsilon(K_i) \setminus K_i \underset{\gamma}{\to} (U_\varepsilon(K_i) \setminus K_i)/\text{sing} \underset{\beta}{\to} U_\varepsilon \setminus z_0\ .$$

This proves lemma 2 and theorem 1. We now also know (summarizing)

Theorem 2. *Every critical component K_i of a height function h_z on a tight immersed surface is intersection of a nested sequence $U_\varepsilon(K_i)$, $\varepsilon \to 0$, of smooth manifolds with boundary Γ, a union of smooth circles. Moreover $U_\varepsilon(K_i) \setminus K_i$ is homeomorphic to $]0,1[\times\Gamma$ and it has by the Gauss map g a cell-like covering onto $U_\varepsilon \setminus \{z\}$.*

Note that $U_\varepsilon(K_i)$ may be any compact orientable surface with boundary.

Definition. *Let M / Sing be the quotient space of M by the minimal equivalence relation that sends any critical height function component K_i which is not simply connected into one point K_i / Sing, for $i = 1, \ldots, v$. Then we have*

Theorem 3. *If $f : M \to \mathbb{R}^3$ is a tight smooth surface immersion, then M / Sing can be (clearly) obtained from w closed surfaces $M_1, \ldots, M_w$, by identifying isolated points on these surfaces into the points K_i / Sing, $i = 1, \ldots, v$. The Gauss-map g restricted to M_u is a cell-like branched covering of P for $u = 1, \ldots, w$.*

Here a cell-like branched covering is by definition the composition of a cell-like map and a branched covering. A branched covering is by definition locally expressed by $\zeta \to \zeta^k$ for $\zeta \in \mathbb{C}$ and $k \geq 1$. Note that the closed surfaces M_u are not embedded in $\mathbb{R}^3$.

Proof. If K_i is simply connected, then K_i is cellular, as it is the intersection of a nested sequence of surfaces with boundary, that have as boundary one circle and are immersed in the tangent plane of K_i. Such surfaces are discs.

If K_i is not simply connected then $U_\varepsilon(K_i)$ / Sing is a disc (as in figure 4) or a wedge of at least two 2-discs attached to each other at a common center. For examples see the figures 1,2,3. So the natural map $M_u \to M_u$ /sing is cell-like and the map M_u / sing $\to P$ is outside Morse-singular points a covering. It becomes a branched covering if we add these points again.

Remark. For *analytic* tight surfaces the conclusions simplify very much, as critical components of height functions then must be components of analytic sets in $\mathbb{R}^2$: point or circle. In [H$_2$] upper and lower bounds are given for the numbers of Morse singular critical components of smooth as well as well as analytic tight surfaces in $\mathbb{R}^3$.

Locally tight surfaces in $\mathbb{R}^3$.

Call a smooth immersion $f : M \to \mathbb{R}^3$ of a closed surface *locally tight* if all non degenerate height functions h_z have the same number $c \geq 4 - \chi(M)$ of critical points. Then we can prove

Theorem 4. *The conclusions of theorems 1,2,3 hold also for locally tight surfaces, except for the possibility of more than one components* M_u *where* $\kappa \geq 0$.

If f is not locally tight, then f is not a critical point of the functional τ. f is locally tight if τ has a local minimum at f.

The double covering of a non orientable tight surface like in figure $2c, 2d$ is locally tight. In figure 5 we show an anlytic locally tight immersed sphere M with a vertical axis of rotation-symmetry. Here $c = 6, M_1$ and M_2 are represented on two convex surfaces.

Isotopy tight knotted embedded surfaces studied in [KM] are also locally tight.

References

[CL] S.S. Chern, and R. Lashof, *On the total curvature of immersed manifolds.* I Am. J. of Math., 79(1957) p. 306-318. II Michigan Math. J., 5(1958) p. 5-12.

[CR] T.E. Cecil, and P.J. Ryan, *Tight and taut immersions of manifolds.* Research notes in mathematics, vol. 107. Pitman Publ. Inc. (1985), 336 p.

[H_1] F. Haab, *Immersions tendues de surfaces dans E^3.* Thèse IMPA, Rio de Janeiro (1990), à paraître dans Commentarii Math. Helvetici.

[H_2] F. Haab, *Surfaces tendues dans E^3 et nombre d'ensembles Morse non singuliers.* En préparation.

[K] N.H. Kuiper, *Geometry in total absolute curvature theory.* Perspectives in Mathematics. Anniversary of Oberwolfach 1984. Birkhaüser p. 377-392.

[KM] N.H. Kuiper, and W.F. Meeks, III. *Total curvature for knotted surfaces.* Invent. Math. 77 (1984), p. 25-69.

[RS] J.H. Roberts, and N.E. Steenrod, *Monotone transformations of two-dimensional manifolds.* Ann. of Math. 39(1938),p.851-861.

F. Haab : Departamento de Matematica, UFMG, 31270 Belo-Horizonte, Brasil.

N.H. Kuiper : Institut des Hautes Etudes Scientifiques, 91440 Bures -sur-Yvette, France. Math. Inst. Universiteit P.O.Box 80010, 3508 TA, Utrecht. Private adr. : Hoenderkamp 20,6666 AN, Heteren, Netherlands.

Tel. : 31-(0)8306-43507

My interaction with S.S. Chern*

J. Simons

I'm the only one who bought my way up to the platform. Everyone else got in for free. I bought myself a slot in this program. Well, it will be a very small slot. I'm delighted to be here and I'm delighted to be with Chern again and all the mathematicians.

Ambrose used to be fond of saying, well he was fond of saying many things, but one thing in particular, he used to say ones life is determined by events of small probability. Such an event was my going out to Berkeley. After Chern left Chicago, he went to Berkeley. I was a graduate student at MIT, having been an undergraduate there too; Ambrose and Singer said, "You know you've been here long enough. You'd better go someplace else; go to Berkeley and learn from Chern." who was about to show up there. That was the year that Chern, Spanier and Smale and somehow everyone converged on Berkeley I think 1959 perhaps. So I went down to Berkeley, got to know Chern and have a study with him. I was a student of Koscant: I learned from

*editor's note:
This was the talk given by Dr. Simons. Based on the videotape of the conference, B. Cheng and L.T. Cheng wrote up this account. We note that Dr. Simons sponsored this conference by covering the travel expenses of several speakers.

Singer, Chern and I got to be sort of friends but that by being out in Berkeley ended up I would say –– I have a granddaughter now in Berkeley, an ex––wife in Berkeley, a son in Berkeley, a daughter in Berkeley - somehow that their getting rid of me there at MIT because Chern was going to Berkeley was very influential.

Even though I didn't really study with Chern, although people assumed that I did. When I was there, he was always very encouraging and later on, when I proved some theorem in Minimal varieties I called him and showed him this theorem. He said, "Oh, global! Global theorem is very good. It's rare to have a nice global theorem." Boy that was very very encouraging. I just wanted to run home and prove a million global theorems. During these few years I was working on Minimal Varieties, when I had something that of being global, I would call him up - another global theorem. He was always glad to get a call from me and Mrs. Chern always acted glad that I was on the phone - I'm not sure she really was.

Later on when I did start this work that Yau talked about which is Chern-Simons Invariants. I showed the initial work to Chern and I was very lucky I did because he understood the much broader implications. He understood how to generalize it to many dimensions and helped me understand and then we could gather what some of the implications really were of this stuff that I had been working on.

Anyone who has ever worked with Chern or interacted with him etc –– he's just a marvelous fellow, he was always totally encouraging, totally supportive. I was thinking of all these adjectives and I ended up writing the Boy Scout Law –– if anyone was a Boy Scout you might remember this law in litany of positive adjectives: trueful and be loyal, helpful, friendly, courteous, kind, obedient, gentle, thrifty, brave, clean and reverant. Now I don't know about reverent and I don't know about obedient. I was never, obviously, in a position to tern that in particular, that characteristic. But pretty much every other of those characteristics is bodied in Professor Chern. He helped me in mathematics and I'll tell you one remark he didn't make to me but that passed through the grapevine back to me which I think illustrated two marvelous characteristics. Some of you know I left mathematics in the mid-70's and continued to not be a mathematician. At the time this was a field where some leads are no use. It's terrible and so on and so far on. Finally he reported back that Chern said, "Well, after all he was no David Hilbert" —— it wasn't David Hilbert, it was somebody else, which in my opinion illustrated two important points, aside from stating the abvious. First of all, he was a philosophical fellow, clearly that science was not going to go down the drain as a result of my leaving —— that illustrated the philosophical. The other thing was the delicacy of his reach. There was certainly others whom I was also known. He didn't have reach as high as David Hilbert so I feel that it illustrated a lovely touch and a good philosophical sense and I'll always be grateful to Chern for having a totally positive impact on my life.

S.S. Chern: Mathematical Influences and Reminiscences*

Manfredo P. do Carmo
IMPA, Rio de Janeiro

In September of 1960 I left Brazil and went to Berkeley to study differential geometry with S.S. Chern. I had made the decision of dedicating myself to mathematics and it was therefore natural to obtain a Doctor's degree at a good university. It is possible that the choice of differential geometry was influenced by my background as an engineer and my interest in questions of physics. At any rate, once differential geometry was chosen it was almost a consequence that I should study with S.S. Chern. Fortunately, Chern already knew some Brazilian mathematicians and through their letters of recomendation I was accepted as his student.

About that time, a group of mathematicians interested in geometry and topology (Chern and Smale included) moved away from Chicago into Berkeley. This was the starting point of an intense activity in these areas that would extend throughout the decade and would transform Berkeley, by the end of the sixties, in the center of gravity of differential

*A modified version, in Portuguese, of this lecture was published in Matemática Universitária, 12 (1990), 1-5

geometry and dynamical systems.

I stayed in Berkeley until January of 1963, when I obtained my degree and returned to Brazil. During those two and a half years I had the privilege of becoming acquainted with a number of mathematicians (Berger, Chern, Klingenberg, Kobayashi, Smale, Spanier, to name only a few) but the strongest influence of this period is definitely that of Chern.

It was through his teachings that I came to a vision of mathematics as a unified science in which there is no sharp distinction among the various areas, and where a controlled intuition plays a fundamental role. It often happened in his classes and seminars that a combination of algebra, geometry and analysis would come up blended in a beautiful way. Appeals to the intuition were cleverly made (so cleverly, in fact, that sometimes they were only noticed when reading back the class notes), and, with a fine sense of humor, he would quote recent results from other areas by saying: "My analyst friends assure me that I can use this fact".

As an advisor, Chern was far from domineering and would leave his students more or less free to pursue their own interests. Actually, he would prefer each student to find his own problem starting from the indications that he would provide. My thesis, for instance, had as a starting point a paper of Rauch that was considered at the time somewhat hard to read. Chern pointed out the paper to me and said: "If you understand this paper, it might give you a thesis". I was lucky that, at that time, two of the experts on the subject, Marcel Berger and Wilhelm Klingenberg, were visiting Berkeley – invited by Chern, by the way. By talking with Berger, I obtained the necessary indications to become familiar with the literature on the subject, and with the help of Klingenberg, I was able to formulate and solve a question that was implicit in Rauch's paper. However, without Chern's encouragement, I would probably not have even started.

This kind of help that Chern would give to his students was of the same type of that he would give to his mathematical colleagues: general suggestions, based on a deep vision of things that were really important. Details should be treated by each one on his own account and risk. I recall that in my first year in Berkeley I was trying to read a paper by Cartan and there was a passage that I could not follow. Since Chern was a well-known expert on Cartan's works, I went to his office and said: "I don't understand this part". He took the volume, leafed through the paper, looked attentively at the passage, and said: "Neither do I". At this moment, somebody half-opened the door of the office, and, seeing us, said somewhat embarassed: "Sorry, I did not want to interrupt". Before he could close back the door, Chern said: "No, no, no; you are not interrupting. We were more or less finishing. Please, come in". I do not remember who was the person or what was it about but after a few moments, Chern turned to me and said: "Manfredo, we will come back to that subject later; in the meantime, you will probably solve the question yourself". I learned the lesson and never again asked him about details.

As I said before, Chern treated his thesis students on the same footing as he treated his mathematical colleagues. This was flattering but might become embarrassing. I remember that as soon as I was convinced that I had obtained the result of my thesis, I run excited to his office to tell the news. There had been a Congress of Differential Geometry and Relativity in Santa Barbara, and Berkeley had many "visitors" that were passing by. While I was explaining what I had done, a geometer came into the office and Chern asked me to start again. After a few of these interruptions, Chern said: "Manfredo, there are so many people in Berkeley interested on the subject that we had better do the following. Let us set an informal lecture for, say, the day after tomorrow, and then you explain your theorem to everybody" Now, I had just obtained a result that would probably need a lot of polishing before being presented in public. Anyway, it was impossible to withdraw. The lecture, that took place two days later, became a heated debate on the validity of some of the arguments that I had used. The general impression (and also mine) was summarized by a single sentence of Chern: "It will be a beautiful thesis, provided it is correct". I spent the following three months working hard on the details of the paper which was finally published in the Annals of Mathematics.

I came back to Berkeley in February of 1967, with a Guggenhein Fellowship, and stayed there until June 1969. Berkeley had become the Meca of differential geometry and dynamical systems, due mainly to the influences of Chern and Smale. This was one of the most productive periods of my professional life and, again, the crucial influence was that of S.S. Chern. In the Winter of 1968, Chern gave a course on minimal submanifolds. One of the goals of the course was to cover, using the method of moving frames, a recent preprint of J. Simons, which, in Chern's view, was to play an important role in future developments. The audience included various mathematicians, such as J. Cheeger, D. Gromoll, S. Kobayashi, F. Warner, and others. I followed the lectures closely and before the end of the course, Chern, Kobayashi and myself were able to solve a question implicit in Simons' paper. That was the starting point of my interest on minimal surfaces that would occupy a good part of my research activity in the following years.

Chern had many friends among Brazilian mathematicians. Besides myself, he had four other Brazilian students: Alexandre Rodrigues, Leo Amaral, João Lucas Barbosa and Plinio Simões. My first student, Keti Tenenblat, went to do her postdoctoral work with Chern, and ended up doing some joint work with him. As in my case, the subject on which she worked with Chern became a permanent and growing interest in her research career.

In June of 1979, there was a large meeting in Berkeley, the famous Chern Symposium, to mark the occasion of Chern's retirement. At that time, I was visiting Berkeley for a year, and both Lucas Barbosa and Keti Tenenblat came specially from Brazil for the symposium. The tree of differential geometry, that was somewhat dried up in the beginning of Chern's career, was now a blooming garden, with vivid colors and varied species. Every participant in the symposium was convinced of the fundamental role that Chern had played in this transformation. At the closing dinner, Louis Auslander, one of the first Chern's students, asked for all Chern's students to stand up. It was an impressive view. But the culminating point was reached when Raoul Bott asked to speak and said, among other things: "Very well. Louis asked for Chern's students to stand up, and so it was done. But I want to say here that, in one way or another, we are all Chern's students". The prolonged applause that followed this statement was the greatest tribute that I ever saw given to a living mathematician.

Riemannian Manifolds: From Curvature to Topology

A brief historical overview

Marcel Berger
Mathématiques
Institut des Hautes Études Scientifiques

It is a great pleasure and an honor to contribute to a book dedicated to S.S. Chern on his eightieth birthday. But before starting this very short historical essay, I want to express my enormous personal admiration for Chern. Not only for his obvious contributions to mathematical science, but also for his human qualities.

Something exceptional about him is the attention he has always paid to other mathematicians, whatever their stature. He was always eager to get acquainted with every young differential geometer who came along, and would then store each of them in his fantastic memory. Whenever and wherever he met one of them again, he was always very quick to go up and address him first, even when he was already engaged in conversation with some more eminent person. This generous trait of character belongs only to the truly great.

First I would like to say what this essay is not. It is not a survey of the subject in the usual sense, an exhaustive report with a large bibliography. On the contrary, it is a whirlwind tour, a survey in the sense of a bird's eye view.

The point I want to make is the following. Although it may look completely natural to the differential geometer of today, the problem of the relations between curvature and topology is not so natural as first it seems. I am not going to offer reasons for this, it will be left to the reader's imagination. As we will see, a very long time elapsed between the foundation of the theory of Riemannian manifolds by Riemann in 1854 and Heinz Hopf's papers in the early thirties, where this apparently naïve question was raised.

This whirlwind policy (sadly enough) supposes some limitations. First, I shall be keeping to the most general question, leaving aside, for example, the case of Kähler manifolds, although they boast – especially with the work of Aubin-Calabi-Yau – a wonderful set of results. Secondly, I will stick to a few basic items for the bibliography. Thirdly, I will not go into too much detail about the very intricate interplay between various authors in the 1930's when the subject was still in its cradle.

I shall also systematically ignore the interaction involved in the historical building of the notion of *manifold*. Although obvious to a lot of mathematicians from Riemann to Élie Cartan, the notion is a subtle one (indeed things became clear and solid only with a basic paper by Whitney in 1936). Nor will I touch on the geometrical properties of a given manifold with given curvature sign. For the specially rich negative case, see for example [1].

Last but not least, I hope to be forgiven by the many important contributors in the field whom I do not cite. And I hope that some colleague will one day write a detailed history of global Riemannian geometry from Gauss to, say, 1980.

The keystone of our text is Hopf's 1932 program. First a very brief mention of definitions and results prior to Hopf. Then some comment on the basic questions of his program. Finally a brief description of Hopf's legacy, in which Chern's contributions play an essential part. However, as will be briefly explained in due course, Chern's pioneering work stemming out of one of Hopf's questions goes far beyond our topic. It is, in any case, fully discussed elsewhere in this book. For detailed texts on our subject, more of the "report" type, see, for example, Gromov [2], Sakai [3], Dombrowski [5].

I would like to thank S. Hildebrandt who invited me to lecture on the history of Riemannian geometry, and W. Ziller who drew my attention to Hopf's articles quite a few years ago.

A first version of this text was looked at by various colleagues who contributed important items, helped me to be more accurate and pointed out various mistakes. It is a pleasure here to cite them: J.-P. Bourguignon, J. Cheeger, J.H. Eschenburg, R. Greene, M. Gromov, K. Grove, J. Lafontaine, C. Margerin and H. Wu.

Note also that this text was written some time ago and that since many new results appeared for which the reader should consult the recent litterature.

1. Gauss, from1784 to 1827

Gauss had a double reason for studying the intrinsic geometry of a surface M^2 set in the 3-dimensional Euclidean space E^3. First he was interested in the foundations of Euclidean plane geometry, more precisely in finding out whether Euclid's fifth postulate was really necessary or not. Secondly, ordnance-type surveying was one of his jobs as director of the Göttingen Astronomical Observatory. A basic question from both points of view was the study of the sum of the angles of a triangle.

Gauss's fundamental discovery can be sketched roughly as follows. If you write the infinitesimal element of length $g = ds^2$ of $M^2 \subset E^3$ and if you take a coordinate chart (x, y) around a point $m = (0,0) \in M^2$ such that the lines through m are geodesics, then the limited expansion necessarily takes the form

$$g(x,y) = dx^2 + dy^2 + K(m)(xdy - ydx)^2 + o(x^2 + y^2) \,. \tag{1}$$

This proves of course that K is an intrinsic invariant of the length metric induced on M^2 by E^3, but Gauss succeeded in proving that

$$K = k_1 k_2$$

where k_1 and k_2 are the principle curvatures of M^2 at the point considered (and of course k_1 and k_2 separately – or their sum – depend not only on g but also on the embedding).

We open a small parenthesis here for the curious reader. No text on the subject answers the question that immediately comes to mind. Namely, can one deform (at least locally) M^2 in E^3 really to change the embedding whilst keeping invariant the induced length metric g? Some texts here give the example of $K \equiv 0$ obtained by any developable ruled surface. But it is an extremely special case. It is indeed true that you can locally deform almost every piece of $M^2 \subset E^3$ whilst keeping the first fundamental form g invariant, but it is hard to prove, and the question is not yet wholly understood. For this see [4], page 427, and the references given there.

Let us comment, in passing, on the case where $K \equiv 0$. It is not hard to see that it characterizes the fact that g is locally isometric to E^2, the Euclidean plane. So in a good sense the function $K : M^2 \to \mathbb{R}$ is a measure of the *deviation* of (M^2, g) from being Euclidean. Gauss's work is extremely well analyzed and put into a modern setting in [68].

2. Riemann 1854

Riemann was interested in the foundation of geometry, and this in quite a general sense. Moreover at that time the hyperbolic geometry of Lobatchevsky and Bolyai was not considered to be set out in a proper mathematical way. Riemann's basic paper (see [5] for a translation and an extremely good commentary) is an amazing piece of pioneering work. He defines, though not very rigorously, an n-dimensional manifold M^n and a "Riemannian" metric on it: $ds^2 = g$ is a positive definite quadratic form on every space tangent to M^n, and depends differentiably on the point. Such a Riemannian manifold $(M = M^n, g)$ has a metric, which is given by

$$d(m, n) = \inf\{\text{length of curves from } m \text{ to } n\}$$

the length of a curve $t \mapsto m(t) \in M$ being of course $\int_{t_0}^{t_1} \sqrt{g[m'(t), m'(t)]}\,dt$. Then (a sparkling example) the hyperbolic plane everybody was looking for is simply given by

$$M^2 = \{(x \cdot y) \in \mathbb{R}^2 : x^2 + y^2 < 4\}$$

$$g = \frac{dx^2 + dy^2}{\left(1 - \frac{x^2+y^2}{4}\right)^2}$$

The extension of this definition to any dimension is trivial.

The major problem now is to find what generalizes the Gauss curvature K (cf. equation (1) above) in the case $n = 2$. In a coordinate system $(x_1, \ldots, x_n)$ where all lines through the origin $m = (0, \ldots, 0)$ are geodesics, Riemann proved that equation (1) becomes

$$g(x_1, \ldots, x_n) = dx_1^2 + \ldots + dx_n^2 + \sum_{1<j,k<h} R_{ijkh}(m)(x_i dx_j - x_j dx_i)$$
$$(x_k dx_h - x_h dx_k) + o(x_1^2 + \cdots + x_n^2) \quad (2)$$

The consequences of the fundamental formula (2) are numerous. Our first remark is a sad one: when $n \geq 3$ there is no really significative function $M^n \to \mathbb{R}$ generalizing the Gauss curvature. One cannot avoid working with the Riemann curvature *tensor* $\{R_{ijkh}\}$. For example a necessary and sufficient condition for (M^n, g) to be locally isometric to the Euclidean space E^n is that $R_{ijkh}(m) = 0$ for every i, j, k, h and m. Such spaces are said to be *flat*.

When one considers R as a 4-linear form $R(x,y,z,t)$ it is antisymmetric in (x,y) and in (z,t) and moreover it verifies

$$\left.\begin{array}{l} R(x,y,z,t) + R(x,z,t,y) + R(x,t,y,z) = 0 \\ R(x,y,z,t) = R(z,t,x,y) \end{array}\right\} \quad \text{for every } x,\ y,\ z,\ t\ . \tag{3}$$

One natural idea is then to consider R as an endomorphism of the second exterior power $\wedge^2 TM$ of the space tangent to M. The induced map $R : \wedge^2 TM \to \wedge^2 TM$ is called the *curvature operator* and denoted by $R(\wedge^2)$.

To get real numbers out of R the best we can do (taking the skewsymmetries into account) is to look at the $R(x,y,x,y)$. When $\{x,y\}$ is an orthonormal pair we write

$$K(x,y) = R(x,y,x,y)$$
$$\left(\text{and more generally } K(x,y) = \frac{R(x,y,x,y)}{\|x \wedge y\|^2}\right)$$

and we call it the *sectional curvature* of the 2-dimensional tangent subspace generated by x and y. So that finally the most geometric invariant attached to an (M^n, g) is the map

$$K : G_2 M \to \mathbb{R}$$

so obtained from the Grassmannian of the tangent 2-planes to M into the real numbers. Note that when g is changed to λg then K changes to $\lambda^{-1} K$. This explains the power $\frac{n}{2}$ which appears in integral variants below.

The symmetry relations (3) easily imply that knowing K is equivalent to knowing R (or $R(\wedge^2)$). But the relations between K and R are subtle and still not completely understood (e.g. what the critical planes of K are, and how they are distributed).

Because we will need it later on, let us say how, in a modern setting, one can prove that $R \equiv 0$ is equivalent to the metric being locally Euclidean.

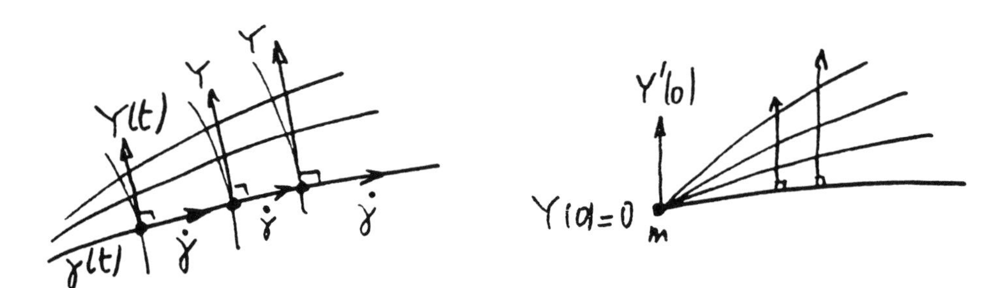

What follows is a vector and global version of (2)

What is a local – and not only second order – version of (2)? Namely, along a geodesic γ through a point m the equation governing the infinitesimal displacement along γ is

$$Y'' + R(Y) = 0 \qquad \text{with} \quad R(Y) = R(\dot\gamma, Y)\dot\gamma \tag{4}$$

where the second derivative is interpreted in the sense of parallel transport along γ and R is the endomorphism dual to the 4-linear form R. This equation shows that if we know R around m and what the parallel transport does to it, we know g locally. In the special case where $R \equiv 0$ it is now easily done.

3. The *Raumformen* Problem from 1854 to the Present

To understand and to classify the so-called *space forms* was one of the main driving forces in the continuation (with the use of more and more solid materials) of the construction started by Riemann.

In the air was the idea that there were two other geometries, besides the Euclidean, that were rich enough, that is, the same for any two-dimensional subspace (one says sometimes "2-points isotropic"). This condition is locally equivalent to K being identically a constant k. Equation (4) shows immediately that locally $k > 0$ yields spherical geometries and $k < 0$ hyperbolic ones. This was understood early on. But the difficulty was the global problem. The first thing was to define a space form, what is know today as a *complete* Riemannian manifold with constant K. The second problem was to classify such forms. It is not because you know that the universal covering is unique when k is given that you have finished the job.

To telescope history let us say this. It was not before the Hopf-Rinow paper of 1931 that the notion of complete Riemannian manifold was correctly stated, and this, moreover, only for $n = 2$. Then, it being accepted that the general dimension is not really different and that the notion of manifold is clear, it was essentially in the thirties that the following became known:

i) every Riemannian manifold of dimension n and with $K \equiv k$ is locally isometric to the Euclidean or the sphere or the hyperbolic space of the same dimension. We denote by Hyp^n the space with $K \equiv -1$.

ii) if the manifold is simply connected and complete, then the above isometry is global

iii) as a consequence of i) and ii), any such complete manifold is a quotient of one of the simply connected (unique) models by a discrete group of isometries acting freely on the universal covering.

Even before this was completely and correctly established, people started to work on the problem of classifying these space forms. The flat case was part of Hilbert's 1900 address and Bieberbach made a basic contribution in 1911: compact space forms have to be finite quotients of flat tori. The classification of compact space forms with $k = 0$ and $k > 0$ was almost completed around 1967, see [6].

The negative case (hyperbolic space forms) is a much more difficult story. First, building explicit examples is not trivial and involves number theory most of the time. For $n = 2$ there are continuous families of non isometric examples; this is Teichmüller's theory, still progressing but well understood. It was only in 1931 that Löbel found infinitely many non-isometric 3-dimensional compact hyperbolic space forms. In the 60's, works of Calabi, André Weil and L. Mostow culminated in a rigidity theorem for any $n \geq 3$, in sharp contrast to Teichmüller's $n = 2$ situation.

One of the latest results is the discovery by Gromov and Piatetski-Shapiro of non-arithmetic compact hyperbolic space forms for any n (see [26]).

For the sake of simplicity we will now assume that – unless explicitly stated otherwise – all manifolds are COMPACT (whence necessarily complete).

4. Heinz Hopf: The 1932 Paper

In the *Jahresbericht der Deutschen Mathematischen Vereinigung 1932* [7], Heinz Hopf published a paper which seems to have been the first to explicitly raise the question of the implications of the sign of K on the topology of a (compact) Riemannian manifold (M^n, g). It will be clear from the following that Hopf was working on the question long before 1932 and also probably mentioned it to many people. The title of his paper was *Differentialgeometrie und topologische Gestalt*. Looking at $n = 2$ he recalls that for space forms (i.e. $K = cst$) one has three properties (we shall be using Hopf's notations below):

B' $K > 0$ implies compactness (completeness is understood)

B'' The sign of the Euler characteristic $\chi(M^2)$ of the space form M^2 is equal to the sign of K

B''' On any given space form the sign of the curvature of a metric of constant curvature is unique.

Then Hopf seeks to generalize this, asking the same questions as above:

P' Is any space form (understood to be complete) of positive sign necessarily compact?

P'' When n is even, has the Euler characteristic the same sign as the constant curvature to the power $\frac{n}{2}$?

P''' Does a space form carry only space form structures of the same sign (understand here that there are three signs: -1, 0, $+1$)?

We will comment on these questions further on. Because Hopf turns immediately to the following three extremely general questions which I quote in the terms used.

"The problem of determining the global structure of a space from its local metric properties and the connected one of metrizing – in the Differentialgeometrie sense – a given topological space, may be worthy of interest for physical reasons." We can specialize this to the Riemannian case, still quite general, in:

Problem I What is the best Riemannian metric on a given compact differentiable manifold?

Problem II What are the implications on the topology of M when certain properties of the curvature of a Riemannian metric g on M?

Problem I is fascinating. It merits an essay in itself and we shall it aside altogether. It is in full progress these days. In this precise form it was mentionned to us by René Thom in the Strasbourg mathematics department library in the sixties. Problem II, rephrased, is none other than "FROM CURVATURE TO TOPOLOGY", the theme of the present article. And the rest of Hopf's paper clearly shows that he had precisely that in mind. For example, *extend* his question P'' to variable curvature K of a given sign.

Let us now briefly recall why B', B'' and B''' above are true.

For B' it was a semi-local theorem of Bonnet that a geodesic of length greater than $\frac{\pi}{\sqrt{k}}$ in a (M^2, g) with $K \geq k > 0$ cannot minimize the length between its extremities. With the correct Hopf-Rinow notion of completeness and the existence of a shortest geodesic between any two points, it follows that $K \geq k > 0$ and "complete" implies that the diameter of M is smaller than or equal to $\frac{\pi}{\sqrt{k}}$, hence that M is compact.

B'' and B''' are deduced in one shot from the Gauss-Bonnet formula

$$2\pi\chi(M^2) = \int_{M^2} K(m)dm \quad \text{(for the proof see below)}$$

Hopf then asks more precise questions.

The first is how to extend to any dimension the fact that $K \geq k > 0$ implies the bound $\frac{\pi}{\sqrt{k}}$ for the diameter (we will come back to this later on).

The second problem is to extend the Gauss-Bonnet formula. He adds, *"One should next make clear which curvature quantity should be taken to replace K"*. And again, *"The problem of computing the topological invariants of a variety from its metric is extraordinarily attractive"*.

5. Hopf's Results

In fact this 1932 paper came five years after [8] and [9], published in 1925, where Hopf succeeded in extending the Gauss-Bonnet formula to any *even* dimension n, first for space forms (this solves P'' and P''' is solved in the appendix of [7]) and then for hypersurfaces of E^n. In the first case ($K \equiv k$) the curvature algebraic object to integrate (up to a scalar depending only on n) was $k^{n/2}$; in the case of the hypersurface it was $k_1 \ldots k_n$ where the k_i are the principal radii of curvature.

Now for a few words about the proofs. The first extended the proof of Gauss-Bonnet obtained by triangulating a surface M^2. The Gauss formula proper holds for a geodesic triangle T and says

$$\int_T K(m)dm = \alpha + \beta + \gamma - \pi$$

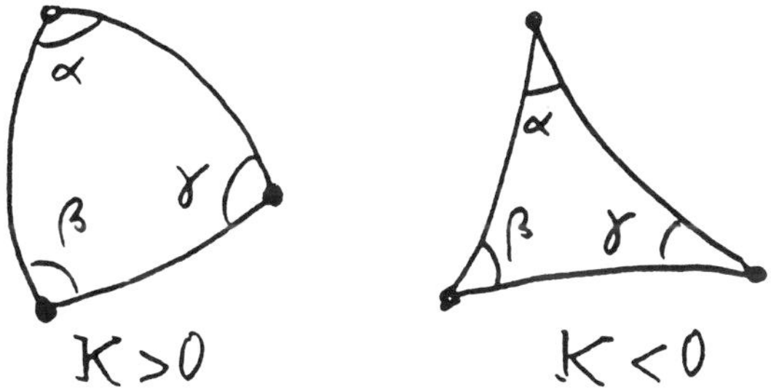

K>0 K<0

where α, β, γ are the angles. Bonnet then completed the proof by summing over the triangles and applying Euler's formula

$$\#(\text{vertices}) - \#(\text{edges}) + \#(\text{triangles}) = \chi(M^2) \ .$$

The extension to simply connected spaces of constant curvature was possible because there exist simplices with well-defined dihedral angles for faces. One then uses a formula of Schläfli (rediscovered by Poincaré in 1905). For variable curvature spaces, and when the dimension is ≥ 3, simplices do not have face angles. A combinatorial proof of the Gauss-Bonnet formula for the variable curvature manifolds of higher dimension is available but requires delicate approximation arguments (see [10]).

The case of the hypersurface is more subtle. Even for surfaces Hopf's proof was new. The cornerstone of the edifice is Hopf's formula computing the Euler characteristic of a manifold M^n as the sum of indices of any vector field on M^n. Then one considers the Gauss map of $M^n \to S^n$, where S^n denotes the unit sphere of E^{n+1} if $M^n \subset E^{n+1}$. The degree of this map is

$$\frac{\int_{M^n} k_1(m) \cdots k_n(m) dm}{\text{volume of } S^n}$$

Hopf then proves a second theorem, namely that the degree of the Gauss map is equal to the sum of the indices of a vector field on M^n. The proof rests on a clever trick, see [14] or [15].

From at least 1925, Hopf relentlessly worked on, or sought for, a formula encompassing for general Riemannian manifolds the two very special cases he had been able to settle. We will come back to this soon. But before that let me mention an important contribution along the road "from curvature to topology".

6. Back to Hadamard 1897-1898

Hadamard had brought out in 1898 a wonderful paper [11] in which he studied the surfaces $M^2 \subset E^3$ with $K \leq 0$, i.e. with the principal curvatures k_1, k_2 of opposite signs. Apart from our own interest in it, this paper is becoming quite famous today in the field of chaos because Hadamard proved sensitivity to critical conditions and the existence of wandering geodesics for a Cantor set of directions at a given point. In any case he proved that the universal covering of what has, since Hopf, been called a complete surface M^2 with $K \leq 0$, is diffeomorphic to $\mathbb{R}^2$. He also mentioned the possibility of extending this result to the 3-dimensional case. The fact that this holds for any dimension and any complete Riemannian manifold with $K \leq 0$ was settled by Élie Cartan in his *Leçons sur la Géométrie des Espaces de Riemann* in 1925. At that time the notion of manifold and that of completeness were still not clearly established, but Élie Cartan always took them for granted.

Because it does not appear to be common knowledge (Preissmann's 1942 paper [12] generally getting the credit) we mention here that Cartan had already proved in 1925 that for a geodesic triangle with sides a, b, c and angle α and facing a in a simply connected M^2 with $K \leq 0$, the following always holds:

$$a^2 \geq b^2 + c^2 - 2bc\cos\alpha\ . \tag{5}$$

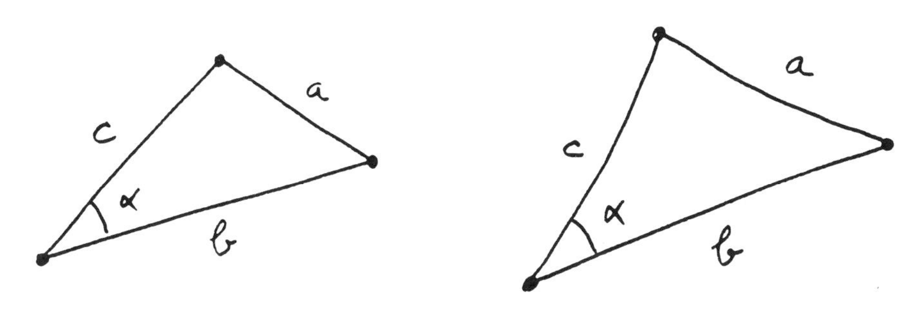

Euclidean case K ≤ 0 case

The proof uses the ordinary differential equation $Y'' + R(Y) = 0$ and boils down only to saying that $K \leq 0$ implies $R(Y) \cdot Y \leq 0$ hence $(\|Y\|^2)'' \geq 0$.

In 1897 Hadamard also worked ([13]) on the case $K > 0$ where he found the study of geodesics to be much more difficult. His paper contains the nice result that a compact surface $M^2 \subset E^3$ with $K > 0$ is the boundary of a convex body in particular diffeomorphic to S^2 even if self-intersections are allowed *a priori*. His proof extends immediately to any dimension n and is too beautiful not to be quoted: $K > 0$ implies that the Gauss map $M^2 \to S^2$ is a covering. But S^2 being simply connected, the Gauss map is now one to one. And so ends the story, with another simple geometric argument.

7. Generalizing the Gauss-Bonnet Formula to any Dimension and to any Manifold and What can be done with it

It was only fifteen years after Hopf's two papers and his persistent questioning that Allendoerfer and Fenchel found independently the curvature quantity that has to be integrated over a (compact) Riemannian manifold to get its Euler characteristic. Not surprisingly, it was hard to guess this quantity, which should in the general case be

$$\begin{cases} k^{n/2} & \text{when } K \text{ is identically a constant } k \\ k_1 \ldots k_n & \text{when} \quad M^n \subset E^{n+1} \,. \end{cases}$$

We know only by a classical formula extending Gauss's $K = k_1 k_2$ that $k_i k_j = R_{ijij} = K(e_i, e_j)$ for an orthonormal basis $\{e_i\}$ and this leads nowhere.

The answer is the universal polynomial of degree $\frac{n}{2}$ in R.... given by the double Pfaffian

$$P_{n/2}(R, \ldots , R) = c(n) \sum \text{ signature } (a_1 \cdots a_n)$$
$$\text{signature } (b_1 \cdots b_n) R_{a_1 a_2 b_1 b_2} \ldots R_{a_{n-1} a_n b_{n-1} b_{n-1}}$$

where $c(n)$ is a constant depending only on n and the summation runs through all permutations of the $\{a_1 \cdots a_n\}$ and of the $\{b_1 \cdots b_n\}$. This formula is especially clear in Allendoerfer's work.

But neither Allendoerfer nor Fenchel guessed the right integrand from the two special Hopf cases: the starting point was a 1939 paper in which H. Weyl succeeded in computing the formula for the volume of a tube of radius ε around a submanifold of E^N of any codimension. So this proof was valid only for Riemannian manifolds which are $M^n \subset E^N$. In 1941 Allendoerfer and A. Weil cleverly succeeded in getting the general case. Which in turn came to be obvious in 1956 wiht Nash's isometric embedding theorem.

In a sense the above proofs are not wholly satisfactory in so far as they are neither intrinsic nor conceptual. One had to wait for S.S. Chern in 1944 before the facts were really understood. Namely that the exterior n-differential form on (M^n, g) (remember that n is necessarily even)

$$P_{n/2}(R, \ldots, R)dm$$

represents, at least in the orientable case, via de Rham's theorem, the Euler class of the tangent bundle of M^n. See the next section for more on this.

But let us now see whether the generalized Gauss-Bonnet formula enables us to answer, extended to variable curvature, Hopf's question P''. It does so in dimension 4 as Milnor has shown (unpublished, see [16]), because, if we take a nice orthonormal basis adapted to K, then $P_2(R, R)$ reduces to $K_{12}K_{34}+K_{13}K_{42}+K_{14}K_{23}+R^2_{1234}+R^2_{1342}+R^2_{1423}$.

But, believe it or not, the extended P'' question is still open today when $n \geq 6$. The reason is that, starting with $n \geq 6$, there exist (local, see [18]) curvature tensors with $K > 0$ and $P_{n/2}(R, \ldots, R) < 0$. This proves that a solution for the extension of P'' to variable curvature – which can be considered as the simplest question in "from curvature to topology"! – should involve not only the generalized Gauss-Bonnet formula, but some global argument that is still lacking, or a counterexample.

What happens is that $P_{n/2}$ starts getting complicated when $n \geq 6$, whilst $P_{4/2} = P_2$ is quadratic and so can be handled. There is a fascinating historical anecdote for this dimension 4 Gauss-Bonnet generalization. It was in some sense discovered by Lanczos in 1938 : he proved that this $\int P_2(R, R)dm$ was independent of the metric, in the sense that its derivative in the space of metrics vanishes identically (this belongs to the important domain of the so-called "null Lagrangian"). Lanczos did not discover or compute that "invariant" because he was trying to extend the Hilbert variational principle for the theory of general relativity. Since everything was vanishing, he considered $\int P_2(R, R)dm$ as *uninteresting* (see [75]).

For example one intrinsic way of writing it is

$$\chi(M) = \frac{1}{8\pi^2} \int_M (\|R\|^2 - \|Ric - \frac{\text{scal}}{4} \cdot g\|^2) dm \,. \tag{6}$$

This formula will not recover Milnor's result but is absolutely basic, e.g. for studying Einstein manifolds, see [17]. But do not try to recover Milnor's result above on $P_2(R, R)$ with integrals on the Grassmannian like $\int_{G_2M} K^2$, $\int_{G_2M} KK^\perp$. It does not work. If $K \geq 0$ or ≤ 0 does not work there is something that does, as discovered by Kostant (unpublished). It is $R(\wedge^2) \geq 0$ (or ≤ 0) as algebra shows without too much trouble. For manifolds enjoying this condition(s) see Section 12.

So, as the title of [18] might suggest, the generalized Gauss-Bonnet formula looks pretty useless when $n \geq 6$. That is true, at least today, if one thinks of the *exact* formula. But its universality has a great theoretical importance (see also the next section). As Cheeger remarked (see Section 22 for more on this) the condition $|K| \leq 1$ and $\chi(M) \neq 0$ implies immediately

$$\text{Vol}(M) \geq a(n) > 0 \,.$$

8. Chern's Breakthrough: Chern Classes and Curvature Integral Formulas for Characteristic Classes

How did Chern prove that the exterior form $\omega = P_{n/2}(R, \ldots, R)dm$ is (under de Rham) the fundamental class of M^n (say orientable and of course still even-dimensional)? He did it by proving that, although not generally exact in M, the inverse image $\pi^*\omega$ in the unit tangent bundle $\pi : UM \to M$ of (M^n, S) is exact. There exists an $(n-1)$-exterior form σ on UM such that $d\sigma = \pi^*\omega$. Then he used Hopf's formula for the index of a vector field and Stokes' formula down on M. Because, when it is never zero on M, a vector field gives a map $M \to UM$. This was in 1944. But then Chern took a leap forward in extending the above scheme. He succeeded in 1946, in the same move, in defining characteristic classes for every complex vector bundle over a manifold and in expressing them by the curvature form of any connection on such a bundle. First, in algebraic topology, this made completely clear the preceding theories of Whitney, Stiefel, Pontryagin. Secondly it gave integral curvature formulas for all characteristic classes. The outcome of this was tremendous and is still not exhausted. Some references are [19], [20], [55], [74], [81].

For our much more restricted topic, we mention first that Cheeger's remark above on manifolds with $|K| \leq 1$ and $\chi(M) \neq 0$ applies in the same way to any manifold with a non-zero characteristic number, i.e. where there exist Pontryagin classes $p_1, \ldots, p_s$ such that their cup-product $p_1 \cup \cdots \cup p_s$ is non-zero when evaluated on the fundamental classes $[M]$ of M. See Section 22.

To temper a little what was said in section 7, it must absolutely be mentioned that the explicit determinant curvature formula for the characteristic classes of any (complex) bundle on a manifold is basic in various recent topics, e.g. secondary classes (see for example [72]. [81], [82]) or various index theories (see for example [19]). We return now to our topic "from curvature to topology" for Riemannian manifolds.

9. The [1935, 1942] Contributions

Here we describe the major contributions to our subject made in that period. As you will see they are sparse and of very different flavors, showing either that the subject was too hard or that it did not interest very many people.

In 1935, Schoenberg and Myers independently solved Hopf's P' question. That is, they proved that the hypothesis $K \geq a > 0$ and completeness did indeed imply a diameter $\leq \frac{\pi}{\sqrt{a}}$, hence, in particular, compactness. Myers added the remark that this implies the finiteness of the fundamental group of the manifold considered. The simplest proof uses equation (4) (already in 1926 Synge had a result in this direction). See also the next section.

In 1936 Synge proved that when one has $K > 0$, even dimension and orientability, then simply connectedness follows (remembering that, from page 7 on, we have assumed compactness for every manifold considered). Synge's trick is wonderful, it uses parallel transport, the second variation formula for the lengths of a one-parameter family of curves neighboring a geodesic and the fact that a rotation in an odd-dimensional Euclidean space always has at least one invariant non-zero vector.

With the exception of those mentioned in the next section, the last result of that period is due to Preissmann. It says that for a manifold with $K < 0$ every Abelian subgroup of the fundamental group is cyclic; for example $\mathbb{Z} + \mathbb{Z}$ is forbidden. His proof consists in rediscovering Elie Cartan's formula (5) (in Section 6) which implies that for a geodesic triangle the sum of the three angles is strictly less than π, something which in dimension 2 follows immediately from the Gauss-Bonnet formula. We conclude with a nice geometric argument: a real parallelogram cannot exist, it has to be squeezed in an interval.

We note in passing that in his 1942 article [12] (only 41 pages), Preissmann managed to give detailed proof of everything that was known before 1942. This shows what slow progress the subject was making.

10. Ricci and Scalar Curvature

In the preceding section an important contribution made by Myers in 1941 was left out. In order to present it, we first need to introduce Ricci curvature. As we have said, the best way to get real numbers from the Riemann curvature tensor, is to introduce the sectional curvature K by specializing x, y, z, t in $R(x,y,z,t)$ to $R(x,y,x,y)$. But there are two other ways (and this time basically no other algebraic ones). The first is to introduce the first trace of R, called the *Ricci curvature*; for tangent vectors x, y we consider the trace with respect to the Riemannian metric g in question:

$$\text{trace}\ (z \mapsto R(x,z,y,z)) = \sum_i R(x,e_i,y,e_i)$$
$$\text{where } \{e_i\} \text{ is an orthonormal basis .}$$

We *denote* this trace by $\text{Ric}(x,y)$; it is a bilinear symmetric differential form on (M,g). The associated quadratic form will be *denoted* also by Ric:

$$\text{Ric}(x) = \text{Ric}(x,x) \ .$$

And if one writes $\text{Ric} \geq a$, it will of course mean that $\text{Ric}(x) \geq a$ for every *unit* vector x.

What Myers proved in 1941 is that the hypothesis $K \geq a$ (for a real positive number a) can be weakened in $\text{Ric} \geq a$ to get compactness (and therefore the finiteness of the fundamental group) via

$$\text{diameter} \leq \frac{\pi}{\sqrt{(n-1)a}}$$

where n is the dimension of M. The proof consists in just taking the trace of $n-1$ equations (4):

$$Y_i'' + R(Y_i) = 0 \qquad (i = 1, \dots, n-1) \ .$$

This was the first example of a topological inference from curvature for a curvature invariant weaker than the sectional curvature. For more on Ricci curvature, besides what follows here below, see [2], pages 70-82 or [17], pages 15-18.

Now for a few words about Ricci curvature. Being a quadratic form it is of the nature of a metric (except that it is not, in general, positive definite). In one sense it is too weak to say much where one thinks the full curvature tensor is required. In another sense, it was Riemann ("wrong and right" intuition) who thought the number of parameters on the metric that was enough to determine the metric – whatever it may mean – was $\frac{n(n-1)}{2}$. We will see in Section 20 that nowadays more and more information can be extracted from the Ricci curvature.

There is of course a weaker invariant, called *scalar curvature*, obtained by tracing the Ricci curvature with respect to g:

$$\mathrm{scal} = \mathrm{trace}_g \, \mathrm{Ric}$$

or, if you prefer:

$$\mathrm{scal} = \sum_{i \neq j} R(e_i, e_j, e_i, e_j) = \sum_{i \neq j} K(e_i, e_j)$$

for any orthonormal basis $\{e_i\}$. This will be the object of Section 21 at the end.

11. Bochner (1946) and his Heritage

Up to now the only techniques we have seen used to discover the topological implications of what we know about curvature were based on second variation for the length of geodesics and the ordinary differential equation (4). It is not surprising that we have results only on the fundamental group $\pi_1(M)$ and hence on the first homology group $H_1(M)$.

A natural idea from today's vantage point is to study the higher dimensional homology (resp. homotopy) group $H_k(M)$ (resp. $\pi_k(M)$) via minimal submanifolds. Then it involves partial differential equations. This is the pendant to Synge's theorem using a periodic geodesic. But because of the dearth of results available on minimal objects in a Riemannian manifold, this was done only recently (see below).

Results on the $H_k(M)$ theory came via the Hodge theory of harmonic forms. According to this theory the group $H^k(M;\mathbb{R})$, isomorphic to $H_k(M;\mathbb{R})$ by Poincaré duality, can be obtained as the set of harmonic exterior differential forms of degree k. These forms ω obey $\Delta\omega = 0$ where the Laplacian-Beltrami operator Δ is elliptic, linear, of second order and intrinsically given by the Riemannian metric g. The basic fact is now the Lichnerowicz formula

$$\frac{1}{2}\Delta(|\omega|^2) = |D\omega|^2 + F_k(R,\omega,\omega) \tag{7}$$

for the numerical function – Laplacian Δ on $|\omega|^2$. In formula (7) the form has to be harmonic, $D\omega$ stands for the covariant derivative of ω and $F_k(R,\omega,\omega)$ is a form which is linear in R and bilinear in ω. For $k = 1$, formula (7) was obtained by Bochner in 1946. The result is then

$$F_1(R,\omega,\omega) = \mathrm{Ric}(\omega^*,\omega^*)$$

where ω^* is the vector field dual of the 1-form ω. This yields at the same time both more and less than Myers' result. For $\mathrm{Ric} > 0$ implies $H^1(M;\mathbb{R}) = 0$, which is weaker than π_1 being finite. But it takes care of the case $\mathrm{Ric} \geq 0$ where this implies that the first Betti number of $b_1(M)$ of M verifies: $b_1(M) = \dim H^1(M;\mathbb{R}) \leq n = \dim M$. Moreover the equality characterizes flat tori.

When $k > 1$ the formula for $F_k(R,\omega,\omega)$ obtained by Bochner looks, even after the Lichnerowicz simplification, complicated and really linked with sectional curvature for example.

This link took some time to understand. Bochner, Lichnerowicz and Yano, among others, obtained intermediate results. In 1953, in particular, they discovered a link between the $F_k(R,\omega,\omega)$ and the curvature *operator* $R(\Lambda^2)$ and proved that the pinching condition

$$\frac{1}{2}Id < R(\Lambda^2) \leq Id$$

implies the positive definiteness of the F_k for any $k = 1,\ldots,n-1$, hence the vanishing of all Betti numbers $b_1(M),\ldots,b_{n-1}(M)$ of M. It took until 1971 and the work of D.Meyer for the positive definiteness of the F_k to be certain so long as $R(\Lambda^2) > 0$. It was then known that only real-homology spheres can bear a Riemannian metric with positive definite operator. We shall see the end of the story further on. But note that the real link between $R(\Lambda^2)$ and the F_k is best understood with spinors, see [20], page 158. Compare also with Kostant's result on the Gauss-Bonnet theorem above.

Just as we have seen the existence of systematic extensions of the Chern curvature formula from the tangent bundle to any vector bundle over a Riemannian manifold (M, g), so there are also extensions of the Bochner technique to generalizations of harmonic objects for some elliptic operators over various vector bundles over (M, g). These are called "vanishing theorems". See [55], [56], the case of the Dirac operator, Section 21 below, and Section 20. Likewise the surveys [59], [60], [61].

12. Rauch's 1950 Paper and its Legacy

H.E. Rauch, a specialist of complex variables – as Chern was to begin with – visited Zürich in 1948. Hopf, still pursuing his drive "from curvature to topology", asked him the following question.

Since we know that a simply connected manifold with K identically a positive constant (say 1 after normalization) is isometric to the standard sphere, would it be true to say that the condition $1 - \varepsilon < K \leq 1$ for a Riemannian manifold (M^n, g) implies that M^n as a manifold is the sphere S^n? Some suitable ε is asked for (depending or not on n). Note that $\varepsilon = \dfrac{3}{4}$ is the best one can expect because on the non spherical compact symmetric spaces of positive curvature and rank 1 the sectional curvature range is exactly $\left[\dfrac{1}{4}, 1\right]$ (more will be said on these at the end of this section). Recall that these simply connected spaces are the complex projective spaces $\mathbb{C}P^m$ of complex dimension m, the quaternionic projective space $\mathbb{H}P^m$ of quaternionic dimension m and the Cayley plane $\mathbb{C}aP^2$. The real dimensions are respectively $2m$, $4m$ and 16. They enjoy a canonical symmetric space Riemannian metric whose curvature range is as above. Moreover their curvature operator $R(\wedge^2)$ is only non-negative but definitely not positive.

The question Hopf put to Rauch can be viewed as a *pinching* question and also as a question of *isolation*. Rauch answered Hopf's question positively in 1952, for $0.68 \leq K \leq 1$ (then independent of the dimension n). His idea brought a completely new vision and technique to Riemannian geometry. Look at the standard sphere, fix a point m on it: then all geodesics starting from m after length π meet at the antipodal point $-m$ and before that they form a nice spray covering the sphere. Technically one says that the exponential map $\exp_m$ at m is a diffeomorphism for the open ball of radius π.

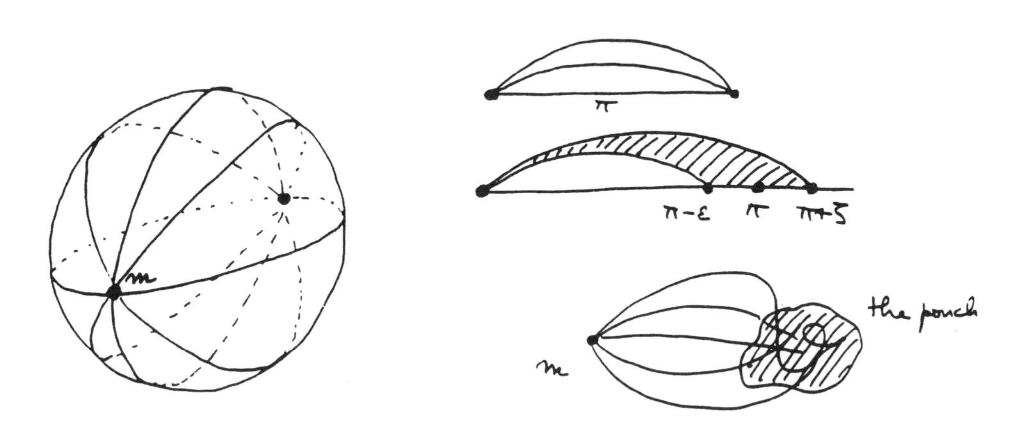

Then Rauch said: if in (M^n, g) the curvature K does not diverge too much from the constant 1, the geodesics emanating from a fixed point m will show behavior similar to that just described for the standard sphere. One expects that up to a length $\pi - \eta$ they will form a nice spray. But from the Myers-Bonnet result above one knows that they have to be stopped before length $\pi + \zeta$ because the diameter

$\leq \pi/\sqrt{1-\varepsilon}$. For ε small enough, $\zeta + \eta$ will be small and what remains of these geodesics between $\pi - \eta$ and $\pi + \zeta$ might be a mess, a *pouch* which one expects to be very small in diameter and then contractible to a point.

In other words one needs to prove that the geodesics from m neither diverge nor converge much more that in the standard sphere. To this end, Rauch invented and proved two *comparison theorems* for manifolds with $a \leq K \leq b$; they are based on equation (4) which, at least theoretically, gives the behavior of geodesics emanating from one point when one takes an initial condition $Y'(0) = 0$ and $\|Y'(0)\| = 1$.

Rauch's proof was quite involved; Karcher remarked in the 70's that for $K \leq b$ there is not much more to be done than to introduce $(\|Y\|^2)''$ and use (4) in the form $Y \cdot Y'' = R(Y) \cdot Y \leq b\|Y\|^2$.

The side $K \geq a$ is harder. Rauch used the Morse index theory, a theory Morse devised to study periodic geodesics. Today Karcher has a new direct proof, see [22].

This was Rauch's pioneering paper. Still, the above presentation – and so the paper itself – did not make completely clear why a small pouch is contractible to a point. This is the story of the *injectivity radius* of a Riemannian manifold (M^n, g). Standard foundations of Riemannian geometry prove easily that for any point $m \in M^n$ there is a number θ such that the exponential map at m on the open ball of radius θ gives in M^n a metric ball of radius θ and diffeomorphic to $\mathbb{R}^n$, hence contractible. For (M^n, g) (remember that all manifolds are compact unless otherwise stated) there is a maximum $\theta > 0$ such that the above is true for every m. It is called the *injectivity radius* of (M^n, g), *denoted* by $\mathrm{Inj}(M, g) = \mathrm{Inj}(g)$. The big problem lies in relating it to the behavior of K, a problem that is still not completely solved today.

The required result on $\mathrm{Inj}(g)$ under $a \leq K \leq b$ was hard to find in Rauch's paper. The notion came to light with a paper by Klingenberg in 1959. Moreover Klingenberg lowered the 0.68 bound given by Rauch to 0.52. For the sake of simplicity we will not follow an exactly chronological order. From Klingenberg we first obtain the dichotomic formula:

$$\operatorname{Inj}(M,g) \geq \inf \left\{ \frac{1}{2} \text{ length of smallest periodic geodesic}, \pi/\sqrt{\sup K} \right\}. \tag{8}$$

The example of a flat torus ($K \equiv 0$) shows that small periodic geodesics cannot be avoided and of course $\operatorname{Inj}(g) \leq \frac{1}{2}$ length of any periodic geodesic.

The dichotomy is not hard because the $\pi/\sqrt{\sup K}$ comes from Myers-Bonnet. To take care of the positive pinching Hopf problem $0 < a \leq K \leq b$, Klingenberg proved two theorems in 1959 and 1961 respectively. The first is:

(i) When n is even, M^n oriented and $K > 0$ then

$$\operatorname{Inj}(M^n, g) \geq \pi/\sqrt{\sup K}$$

(ii) For any n, M^n simply connected and $\frac{1}{4} \leq K \leq 1$ then

$$\operatorname{Inj} \geq \pi$$

The proof of *(i)* is a nice geometric trick. The proof of *(ii)* uses Morse index theory and was in fact given by Klingenberg only with $K > \frac{1}{4}$. The case $K \geq \frac{1}{4}$ was cleared by Cheeger and Gromoll in 1972 but published only quite recently: [21].

To finish the story of the positive pinching problem one needs another tool, the *comparison theorem for triangles*, begun by Alexandrov with surfaces and completed by Toponogov in 1959 for any n. For precise statements and proofs see [22]. The essence of it is that, *without any other condition*, the hypothesis $K \geq a$ (whatever the sign of a) implies that any geodesic triangle has larger angles than the respective angles of the geodesic triangle drawn in the simply connected *Raumform*of curvature precisely equal to a. If we put the above tools together it is not hard to get the so-called "sphere theorem": any simply connected Riemann manifold with $\frac{1}{4} < K \leq 1$ is homeomorphic to the sphere. But the best proof is that of Gromov in [23], see also [2]. It consists in pursuing the pouch idea of Rauch with a basic positive curvature contraction lemma. We have seen above that $\frac{1}{4}$ is the best possible. We finish the sphere case with three points.

The first is the diffeomorphic question. It is still open today; various people (starting in the late 60's with Calabi, Gromoll, Shikata, Sugimoto and Shiohama) have gotten results on it for various ε. The best to date is E. Ruh's, 0.66 valid for any n. But $\frac{1}{4}$ is not excluded. And there exist exotic spheres with no metric of positive scalar curvature (see Section 21).

The second is: what is the curvature condition characterizing a sphere? We have seen Meyer 1971 to the effect that $R(\wedge^2) > 0$ yields a rational homology sphere. It was left to Micallef and Moore in 1987 to prove that $R(\wedge^2) > 0$, and simply-connectedness implies that the manifold is really S^n. The method of proof is very important: it really works by proving that the homotopy groups $\pi_k(M)$ vanish up to $k \leq n-1$, using minimal objects (see the beginning of Section 11). Precisely one looks for energy-minimizing maps, so called *harmonic maps*. These maps are becoming more and more important in geometry; they generalize closed geodesics and, in dimension 2, correspond to minimal objects. The first proof of this kind was the 1980 Siu and Yau proof of the Andreotti-Frankel conjecture to the effect that a Kähler manifold with $K > 0$ is necessarily $\mathbb{C}P^n$ (see [24] for all the above).

The third is the case $\frac{1}{4} \leq K \leq 1$ where M is not a sphere. It is completely taken care of by the so-called "rigidity theorem" (due to M. Berger, see [49]). In this case (M^n, g), if simply connected, is necessarily *isometric* to one of the symmetric spaces $\mathbb{C}P^m$, $\mathbb{H}^q P$, $\mathbb{C}aP^2$. These spaces are the positive "generalized space forms". With their negative pendants of Section 14, spheres and Euclidean spaces, they are the only spaces that are 2-point transitive or, equivalently, isotropic on directions. The word rigidity came from the fact – implied by this result – that on the manifold $\mathbb{C}P^m$, $\mathbb{H}^q P$, $\mathbb{C}aP^2$ one cannot find a Riemann metric g with $\frac{1}{4} < K \leq 1$. Their natural pinching cannot be improved. See also Section 17 for the case "just under" $\frac{1}{4}$.

13. The Other Two Pinching Problems

The condition $1 - \varepsilon < K \leq 1$ can be called the *positive* pinching question. The *negative* one consists in looking for manifolds with $-1 \leq K \leq -1+\varepsilon$. Our guess is that, for ε small enough, the underlying manifold M^n should be homeomorphic (and might even be diffeomorphic) to one of the known ones with $K \equiv -1$, namely a compact space form, quotient of Hyp^n by a suitable discrete group of isometries. The question was solved in the negative by Gromov and Thurston in 1987 when they found, for any $n \geq 4$ and any $\varepsilon > 0$, examples of M^n with $-1 \leq K \leq -1 + \varepsilon$ which are never homeomorphic to a negative space form (see [25], and [76] for a related result).

The *0-pinching* problem is about Riemannian manifolds with $-\varepsilon \leq K \leq \varepsilon$ i.e. $|K| \leq \varepsilon$. Scaling shows that some normalization is clearly needed here. We choose a diameter ≤ 1. Here also our guess is that such manifolds M^n are homeomorphic (and might be diffeomorphic) to one bearing a flat ($K \equiv 0$) metric, see Section 3. And again our guess is wrong. This time also there are manifolds never bearing a flat metric but with a Riemannian metric satisfying $|K| \leq \varepsilon$ and with diameter ≤ 1 for every $\varepsilon > 0$. Their construction is not too difficult: take any manifold of dimension n which can be written as a succession of n fibrations by the circle S^1 finally over a point. Such manifolds are also called *nilpotent* because they can be obtained as quotients by a discrete group of a nilpotent Lie group. An easy computation shows they can be ε – pinched for any $\varepsilon > 0$, always keeping the diameter ≤ 1. The 0 – pinching problem was almost laid to rest by Gromov in 1978 when he proved that for any dimension n there exists an $\varepsilon(n) > 0$ such that an M^n which carries a Riemannian metric g with $|K| < \varepsilon(n)$ and diameter ≤ 1 *has* to be a nilpotent manifold. In fact Gromov still left open the possibility of a finite quotient. This was taken up by Ruh in 1982, see [25] and the references there.

Ideas here are deep and very geometrical: refinements of the comparison theorem for triangles, study of parallel transport along periodic geodesics and the so-called "Margulis lemma" to achieve nilpotency of the fundamental group.

The starting idea, though extremely expensive to work out, consists in looking at the exponential mapping at some point m. There is a ball of radius $\pi/\sqrt{\varepsilon}$ in the tangent plane at m on which the exponential is everywhere of maximal rank. This was Rauch's starting point. When ε is very small and the diameter still ≤ 1 then the counter images of m in the tangent space T_mM at m will hopefully mimic the fundamental group.

14. The $K \leq 0$ Case

As we study "from curvature to topology" a natural question arises: the classification of topological manifolds bearing a Riemannian metric with $K < 0$ (resp. $\leq$, $>$, $\geq$). The condition $K \leq 0$ has a very special nature, since, as we saw in Section 6, the universal covering of the underlying manifold M^n is diffeomorphic to $\mathbb{R}^n$. It looks as though it is all wrapped up ! Up to the classification of the possible fundamental groups...

But remember (Section 3) that the case $K \equiv -1$ is already extremely hard and still not completely understood. The subject is in full progress. Once again, sadly, we cannot treat the fascinating subject, so rich today, of the geometry on a given $K \leq 0$ manifold, see [1]. We just mention a few result, directly related to our subject "from curvature to topology".

But we have the right to wonder about compact manifolds that are the negative pendants of the $\mathbb{C}P^m$, $\mathbb{H}P^q$, $\mathbb{C}aP^2$ of Section 12. Their universal covering is by definition the symmetric spaces of rank one, different from Hyp^n, denoted here by $\mathbb{C}P^m_{<0}$, $\mathbb{H}P^q_{<0}$, $\mathbb{C}aP^2_{<0}$. They admit compact quotients, with of course $-1 \leq K \leq -\frac{1}{4}$.

By Gromov-Thurston examples in the preceding section, one cannot today classify (as in the positive case) the compact Riemannian manifolds with $-1 \leq K \leq -\frac{1}{4}$. But there are almost complete results on the structures of manifold varifying $-1 \leq K \leq -\frac{1}{4}$ on a manifold which is obtained from the beginning as a compact quotient of one of the three $\mathbb{C}P^m_{<0}$, $\mathbb{H}P^q_{<0}$, $\mathbb{C}aP^2_{<0}$ by isometries.

For $\mathbb{H}P^q_{<0}$ and $\mathbb{C}aP^2_{<0}$ we know that there is no Riemannian metric on them wit a strict pinching $-1 < K \leq -\frac{1}{4}$. This was proved in 1990 by Gromov [27], using a sophisticated extension of the technique of the results below on $\mathbb{C}P^m_{<0}$. The idea is that $\mathbb{H}P^q$ and $\mathbb{C}aP^2$ (both > 0 and < 0) admit, at the bundle level, foliations of the type of the Kähler case $\mathbb{C}P^m$. More: in 1990 Gromov and R. Schoen proved that every compact quotient of $\mathbb{H}P^q$ and $\mathbb{C}aP^2$ has to be *arithmetic*, as were all those already known. Arithmeticity is when the discrete subgroup used to make the quotient is (up to a finite extension) contained in the group deduced from the total isometry group of $\mathbb{H}P^q$ (resp. $\mathbb{C}aP^2$) by reducing the real number fields to the integers. The technique is that of non linear partial differential equations, more precisely harmonic mappings, but extended to the case where the target is not a manifold but a Bruhat-Tits construction. Harmonic mappings, even before that, began to be more and more important in Riemannian geometry. In particular they were used by Gromov and Schoen in [84] to prove rigidity for the compact quotients of $\mathbb{C}P^m_{<0}$. The complete story involves works by Siu, Sampson, M. Ville, Hernandez, see [26], [28]. But here the arithmetic question is partially open for the complex case: in 1980 Mostow and Siu found non-arithmetic quotients of $\mathbb{C}P^m$ but only for $n = 2, 3$. In some of the results Bochner's technique (see Section 11), applied to suitable harmonic objects, is basic in the complex case, see [26] and the references there. Recently Corlette found a quaternionic analogue, see [27].

15. The $K \geq 0$ Cases

Constructing examples of manifolds with $K > 0$ is never trivial. In the < 0 case one of the keystones was the theory of discrete subgroups of Lie groups. Another was the Gromov-Thurston technique (see Section 13) of building $-1 \leq K \leq -1 - \varepsilon$ manifolds. The technique involves gluing and blowing-up; this can be done whilst preserving curvature control.

In the positive case one major difficulty is the fact that one cannot (at least today) perform any kind of surgery on a connected sum with a smoothing preserving the positivity of K (unless one is only concerned with scalar curvature, see Section 21). So it took quite a while to obtain the various kinds of examples needed to make an incursion into the still largely unexplored set of manifolds with $K > 0$ or $K \geq 0$.

On our way "from curvature to topology" we will just mention a few of these examples. They are obtained by the Lie group technique, including (non-linear) representations. First, we now know all about compact Lie group homogeneous spaces G/H with $K > 0$ (see [83]). There are only a few exceptions besides Hyp^n, $\mathbb{C}P^m$, $\mathbb{H}P^m$, $\mathbb{C}aP^2$. On the contrary, for a G/H, where G is Lie compact and the quotient metric on G comes from a bi-invariant metric on G, then always $K \geq 0$. It is important to note that there exists (in dimension 7) an infinite family of (homogeneous) manifolds havig different homotopy types. By Weinstein's result (below), this implies that, there, the injectivity radius is not bounded from below even when obviously scaled.

A general remark is that every known example of nonnegative curvature is gotten through Lie group technic, either as homogeneous spaces either with *cohomogeneity* technic.

Now we depart from our "any manifold is compact" assumption to mention the amazing result of Gromoll and Meyer in 1969, see [49]: every *non-compact (but complete) manifold with $K > 0$ is diffeomorphic to R^n*. This was completed by J. Cheeger and Gromoll who proved in 1972 that: *every non-compact Riemannian manifold with $K \geq 0$ has the topology of a vector bundle over a compact Riemannian manifold with $K \geq 0$* (in fact the complete bundle result was obtained by A. Poor in 1974, see [57] for a modern proof). So classification reduces the story to the compact case. The tools involved in the above proof are very geometric, essentially the triangle comparison theorem, but quite elaborate.

So we really have to return to the more direct question: what can be said of a manifold M^n which admits a Riemannian metric with $K > 0$? Note that today we have no idea of exactly which set of manifolds admits a Riemannian structure with $K > 0$ or $K \geq 0$ (not every manifold can do it, see Gromov's result below). Another open conjecture of Hopf's is: can any $M^p \times N^q$ $(p \geq 2, q \geq 2)$ bear or not a metric with $K > 0$; it starts with $S^2 \times S^2$! So in despair and with the pinched case in mind, the first result to mention is:

> *for n even and for a given $a > 0$ there exists only finite homotopy types of M^n with $a \leq K \leq 1$.*

This result is in Cheeger's unpublished dissertation (1966) and was found independently by Weinstein in 1967 ([65]). The idea is worth giving in some detail because it is the inspiration of subsequent results and is moreover quite simple. Besides, with Cheeger's independent result of the same year ([65], see next section), we have a whole new turn of mind, the beginning of a new era in our theme.

Let us call $B(m, r)$ the metric ball of center m and radius r in our (M^n, g). When $r < \text{Inj}(g)$ our ball is contractible. But it is not necessarily so for an intersection like $B(m, r) \cap B(p, r)$. This would be ensured automatically if the $B(m, r)$ were all *convex* for r small enough. Thinking *à la Klingenberg* (see Section 12) it will be true when n is even and $1 \geq K > 0$ as soon as $r < \frac{\pi}{2}$. So now take an economical covering of M^n by $B(m_i, r)$ $(i = 1, \ldots, N)$, i.e. all $B(m_i, \frac{\pi}{4})$ are disjoint and there are as many as possible. Since the sectional curvature permits one (by the Rauch comparison theorems) to control easily both from below and from above the volume of balls of given radius, in particular $\text{vol}(B(m_i, \frac{\pi}{4}))$ and $\text{vol}(M^n, g) = \text{vol}B(m, \text{diameter}(g))$, and since by Myers-Bonnet theorem, $\text{diameter}(g) \leq \frac{\pi}{\sqrt{a}}$, we finally succeed in covering M^n with $N(n, a)$ convex balls. So topology tells us immediately that the possible number of homotopy types for M^n is bounded by $2^{N(n,a)}$.

Since $N(n, a)$ grows like $a^{-n/2}$ when a goes to zero, this method does not give a finiteness result for $K \geq 0$ manifolds. Also it is limited to n even. So for $K \geq 0$ everything seems possible, which is effectively the case with the examples above.

But for the *homology* type we have been safe since 1981, thanks to a result of Gromov's: when $K \geq 0$ the sum of the Betti number obeys $\sum_{i=1}^{n} b_i(M^n) \leq 2^{2^n}$. The tools are the triangle comparison theorem and a new technique of 1977 [37], due to K. Grove - K. Shiohama, making it possible to "go beyond the injectivity radius", using the notion of critical point for the distance function to a given point, defined geometrically with the set of the geodesics joining two points. Moreover this result applies to the Betti number for any field of definition. In fact Gromov's method enabled him (even more simply) to prove that $\sum_i b_i$ can be bounded universally only in inf K (whatever its sign) and the diameter (see [38]).)

This technique of using the *distance function* and its *critical* points should not be underestimated. There are so few "techniques" in modern Geometry! See the survey [67].

Summing up, we have *finiteness* for Betti's numbers of Riemannian manifolds under the self-explicatory notation $\{K \geq \bullet, \text{diameter} \leq \bullet\}$. Which leads us to the next section.

16. The First Finiteness Theorems

In the preceding section we saw an example to the effect that, if one knows the dimension and two positive bounds for K, one can get an infinite number of *homotopy* types. The root of it, as one might expect from the proof of Weinstein, is the fact that the injectivity radius cannot be controlled. In 1967, the same year as Weinstein, J. Cheeger obtained two results. The first is that the set $\{|K| \leq \bullet, \text{vol} \geq \bullet, \text{diameter} \leq \bullet\}$ of Riemannian manifolds permits only a finite number not only of homotopy types but also of *diffeomorphisms*. For the sake of concision, neither here nor later on are we quite exact. Let us just say that in some earlier statements $n = 4$ had to be excluded because differential topology results had to be used to pass from the homotopy to the diffeomorphism type. Today this is no longer needed.

Cheeger's second result is purely geometric, based only on volume and triangle comparison. It says that under the two (three!) conditions $|K| \leq \bullet$ (or $\bullet \leq K \leq \bullet$) and diameter $\leq \bullet$, the two conditions Inj $\geq \bullet$ and vol $\geq \bullet$ are equivalent. One direction is trivial, the other is a nice trick, called "the butterfly". A simplified "cylinder" proof can be found in [30].

A few remarks are now in order. The first concerns Cheeger's technique. He worked by contradiction, assuming there are infinitely many manifolds with different homeomorphism types. If one covers two different Riemannian manifolds (M, g), (N, h) by nice balls (here the hypothesis on the injectivity radius comes in) the infiniteness assumption, together with the curvature hypothesis, makes it possible to control the difference between two of these balls and one ends up with a homeomorphism.

The second remark is that Cheeger's proof can be adapted to give for this finite number an explicit estimate depending only on $\sup |K|$, inf(vol), sup (diameter). See below.

The third remark was announced in Section 8. Namely that the hypothesis vol $\geq \bullet$ (or equivalently Inj $\geq \bullet$) can be dropped when one works with manifolds having a non-zero characteristic number. We will come back to this important remark of Cheeger's in Section 22.

17. Compactness Theorems

Almost explicit in Cheeger's unpublished 1967 dissertation, explicit in his unpublished Stanford lecture of 1970, there is a compactness theorem. It was also implicit in Gromov's article [31] page 283. Things appeared in black and white only in 1981 in the "little green book" [32]. Today everything in the following general plan looks so natural. Consider the set $\mathcal{RS}^n$ of all Riemannian manifolds (compact and, for simplicity, of a given dimension n), where isometric Riemannian manifolds are identified, objects we can call Riemannian *structures*. Then topologize it and look for converging sequences, compact subsets, etc.

In such a framework a result like positive pinching (homeomorphic as well as diffeomorphic) will appear as a properness property of the map from $\mathcal{RS}^n$ into the discrete set $\mathcal{M}^n$ of n-dimensional manifolds, this when looking at the sphere $S^n \in \mathcal{M}^n$. By the same token one can solve the question left open at the end of section 12, namely, "What are the manifolds – different from spheres – which admit a near to $\frac{1}{4}$ pinching, $\frac{1}{4} - \varepsilon < K \leq 1$?" The answer is (at least in the simply connected case) that only $\mathbb{C}P^m$, $\mathbb{H}P^m$, $\mathbb{C}a^2$ do so. This will easily follow from the compactness theorem below. But note that the proof will be non geometrically constructive hence the fact that $\varepsilon(n) > 0$ is not explicit in n.

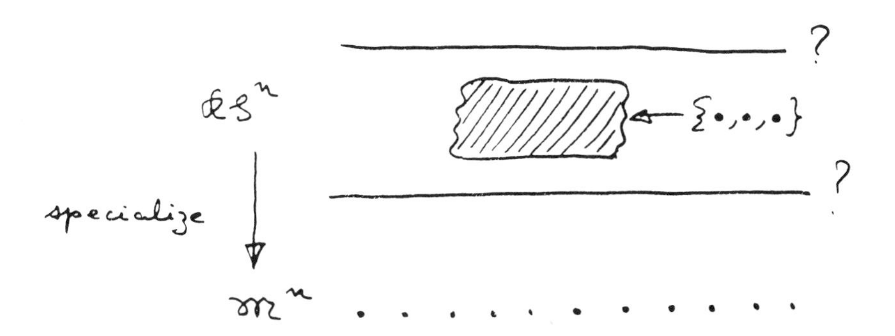

The above compactness result appeared in successively improved versions in the sequence [64], [33], [34], [35]. It reads in [64] or in [33] as: in the subset $\{|K| \leq \bullet, \text{vol} \geq \bullet, \text{diameter} \leq \bullet\} \subset \mathcal{RS}^n$ from every sequence one can extract a sequence converging to an (M^n, g) where $M^n \in \mathcal{M}^n$ and g is a Riemannian metric of (only) class $C^{1+\alpha}$ with $0 < \alpha < 1$. So that we have only *stricto sensu* a precompactness result. It is interesting to note (compare with Section 23) that *harmonic* coordinate charts were used as a basic tool instead of normal (geodesic) coordinates.

From this compactness many of the preceding results could be, at least theoretically, understood. For example when one bounds $|K|$ and the diameter from above and bounds volume from below, then bounds can be deduced on any reasonable Riemannian invariants e.g.

eigenvalues of the Laplacian
length of periodic geodesics
isoperimetric constants and systoles
characteristic and Betti numbers
η-invariant
etc.

Systoles are the lower bounds of volumes (area, length) of various non-homotopic (-homologic) to zero submanifolds (see [73]).

But of course the real game is to devise geometric methods yielding bounds which are explicit in terms of $\sup |K|$, sup (diameter), inf (vol). Examples are the Myers, Bishop and Klingenberg theorems. But in some sense the subject is – except for our particular topic – all global Riemannian geometry, for which we refer the reader to the contemporary literature. Recent references are [19], [69], [70], [71], stopping unfortunately in 1983.

Finally we note that Peters was able to get a (huge) bound for the numbers of possible diffeomorphism type involving explicitly $\sup |K|$, inf (vol), sup (diameter): see [34] for more reasonable bounds. Another remark concerns applications of the compactness result to various pinching problems, such as the "under $\frac{1}{4}$" already mentioned, and also diameter-pinching and volume-pinching, see nice results in [77], [78], [33], [80].

18. The Structure of $\mathcal{RS}$

To understand the implications from curvature to topology in the spirit of the preceding section, we have to study the general sets $\mathcal{RS}^n$ more or less together with various subsets, the first, in historical order, being the $\{|K| \leq \bullet, \mathrm{vol} \geq \bullet, \mathrm{diameter} \leq \bullet\}$.

After quite a gap from Cheeger in 1970 up to Gromov in 1981, the subject became extremely active and today important contributions still often appear. We will look at a few of the basic topics.

First what are reasonable topologies on $\mathcal{RS}$? Back in 1966 Shikata had already introduced a Lipschitz distance. With it he solved the differentiable pinching problem, see Section 12. Gromov's major step was to introduce a Hausdorff distance, valid for the category of more general metric spaces. The fact that they coincide on subsets $\{|K| \leq \bullet, \mathrm{vol} \geq \bullet$, $\mathrm{diameter} \leq \bullet\}$ is crucial. Two kinds of questions arise now.

First: since easy examples show that subsets $\{|K| \leq \bullet, \mathrm{vol} \geq \bullet$, $\mathrm{diameter} \leq \bullet\}$ are not closed when C^2-Riemannian metrics are required, we have to study the closure of these sets in the category of metric spaces. This problem was solved very recently in [35].

Second: what is, more generally, the closure of an $\mathcal{RS}^n$ in the category of metric spaces? It turns out to be basically linked with the topic "from curvature to local topology" as follows. The example of flat tori shows that the injectivity radius can be as small as one likes with $K \equiv 0$, diameter $\leq \bullet$. So it would be a good idea to study the local structure of Riemannian manifolds on balls of radius such as $(\sup |K|)^{-1}$. This was achieved recently by Fukaya in [43], [66]. It is intimately connected, as we have said, with the collapsing of Riemannian manifolds. Collapsing is interesting in itself for describing that part of the boundary of $\mathcal{RS}^n$ which is of smaller dimension as well as the behavior of different invariants under collapses, see [36]. For example the behavior of various characteristic numbers in relation to Chern's formulas expressing them with the curvature. One connected problem is that even when one is working in the realm of compact manifolds, their limits could be non compact.

19. What is Really Needed to Keep a Control on $\mathcal{RS}^n$?

The preceding sections show satisfactory understanding of topology when, from knowing the sectional curvature K, we obtain an upper bound and a lower bound with two extra conditions, one on the volume (or on the injectivity radius) and one on the diameter. These two extra conditions are necessary, as the examples show. But are we not asking too much from the curvature? Of course that depends on what is wanted. Thinking of the result of Myers and Bochner on Ricci curvature (Section 9) and of Gromov on Betti numbers (Section 15), we see that two questions come naturally and are immediately to mind:

- first: does one really need an upper and a lower bound?
- second: can one replace K by Ricci curvature? (for results involving only the scalar curvature see Section 21)

Some other sets of questions will be raised further on. Answers to those above are getting better and better almost every month now. So we have to make a choice among the answers and ask the contributors who are not mentioned to pardon us.

In 1988 Grove-Petersen proved ([41]) a finiteness theorem for $\{K \geq \bullet, \mathrm{vol} \geq \bullet, \mathrm{diameter} \leq \bullet\}$ to the effect that $\sup K$ is not really needed. The geometric idea is the same as that of the initial Grove-Shiohama paper [37]: it is possible to cancel the bound $\sup K$, i.e. to go beyond the radius sup K by using the notion of critical point for the distance function (think of Klingenberg's inequality (8) in Section 12).

We remark that Rauch comparison theorems imply vol $\leq \bullet$ from $K \geq \bullet$ and diameter $\leq \bullet$. This can be contrasted with the following strong result, in 1990 Grove-Petersen-Wu ([42]) found that there is finiteness for $\{\text{inj} \geq \bullet, \text{vol} \leq \bullet\}$.

Note first that the above finiteness theorems are for diffeomorphism types, except when $n = 3, 4$ where for $n = 3$ (resp. 4) homotopy (resp. homeomorphism) types can be obtained. Secondly the first theorem above yielded explicit bounds for the number of manifolds concerned. For more details, we refer the reader to article [34].

20. Here Ricci Curvature Comes In

The most important landmarks following Myer's result $\{\text{Ricci} \geq a > 0\} \Rightarrow \text{diameter} \leq \pi/\sqrt{(n-1)a}\}$, quoted in Section 9, and before more recent results, are Calabi's extension of the maximum principle back in 1957, Bishop's theorem in 1963, Milnor's relation between Ricci curvature and π_1 in 1968, the Cheeger-Gromoll splitting theorem in 1971 (this latter concerns non-compact manifolds and uses partial differential equations) and Gromov's theorem in 1981.

In 1963 Bishop discovered that the volume of balls in a Riemannian manifold (M, g) with $\text{Ric} \geq (n-1)k$ (here k can have *any* sign) grows with respect to their radius no faster than the volume of the balls of the same radius in the simply connected space form $SF^n(k)$ of constant sectional curvature $K \equiv k$. In fact we have more. Consider balls $B(m, r)$ in (M, g) with (any) fixed center m and radius r. Remark that in the simply connected space form of constant curvature k the volume $\text{vol}(B_k(p, r))$ depends only on r, not on p. Then Bishop's statement is:

$$r \mapsto \frac{\text{vol}(B(m, r))}{\text{vol}(B_k(., r))} \text{ is non increasing .}$$

As for Myers' theorem, the proof starts with using $n-1$ equations (4) of Section 2 and taking the trace. There is hope, because a volume is computed as a determinant and because the derivative of a determinant is a trace. But the Ricci curvature is the trace of the sectional curvature!

The simpler bound $\mathrm{vol}(B(m,r)) \geq \mathrm{vol}(B_k(.,r))$ was used by Milnor in 1968 to show that the fundamental group of an (M,g) with $\mathrm{Ric} \geq 0$ has polynomial growth.

In the 1981 "little green book" [32] Gromov used the full strength of Bishop's result – i.e. the radius non-increasing ratio function – to deduce the fact that in the set of metric spaces with Hausdorff topology (see Section 18) the set of Riemannian manifolds with $\mathrm{Ric} \geq (n-1)k$ is precompact.

This raised big hopes of having control on the topology from Ricci curvature bounds alone instead of sectional curvature bounds. Let me say that one cannot expect to be as well off as with sectional curvature, because there is a set of various counterexamples. For manifolds with $\mathrm{Ric} \geq \bullet$ there is nothing here like the Hadamard-Cartan theorem (Section 6): according to Gao and Yau [62] any manifold of dimension 3 admits a Riemannian metric with $\mathrm{Ric} < 0$. More: there is a strong belief among experts that this holds for any dimension. See also [79].

On the postive side, or more generally for $\mathrm{Ric} \geq \bullet$ (any sign), there is no hope replacing $K \geq \bullet$ in Gromov's Betti number theorem of Section 15. In fact a recent construction of Sha and Yang, refined by Anderson, yields manifolds with $\mathrm{Ric} > 0$ and arbitrarily large Betti numbers [63].

Still, results similar to those of the preceding section were obtained very recently by Anderson and Cheeger (to appear). They proved a diffeomorphism finiteness theorem for sets $\{\mathrm{Ric} \geq \bullet, \mathrm{Inj} \geq \bullet, \text{diameter} \leq \bullet\}$ in $\mathcal{RS}^n$ (any n). Moreover there is precompactness in the C^α-topology for any $\alpha < 1$. Finally, counter-examples show that it is impossible to replace the lower bound on the injectivity radius by one on the volume (compare with Section 16).

Here the methods of proof are more sophisticated than those used in the preceding sections, i.e. the triangle comparison theorem, Bishop's theorem and harmonic coordinates. The problem is to control the lower bound for the radius of the elements of the harmonic atlases. Then the notion of harmonic radius comes in naturally. The authors do a great deal of analysis, based on Sobolev inequalities as applied to function on harmonic charts, in particular to the gradient of the distance function for which the Bochner type formula (see Section 11) involves precisely Ricci curvature.

Very recently a major breakthrough appeared in Ricci curvature assumption in the paper [85] of Abresch and Gromoll. It consists in a weak Toponogov's triangle theorem, namely a metric result and not only a volume one, see page 357 of [85]. Many new results are coming using this technic called the *excess* function..

21. What Scalar Curvature Can Yield Today

As for the sectional curvature K and the Ricci curvature Ric we test our question on the case with a given sign. The reader will not be surprised to hear that, well before the Gao and Yau result, Eliasson and Aubin in 1972 proved independently in 1970-1 that on any compact manifold of dimension ≥ 3 there exists a Riemannian metric with constant negative scalar curvature. This is not surprising, the scalar curvature looks like a very weak invariant. To support this, the scalar curvature gives only the second order term in the limited expansion for the volume of balls, the first one being the Euclidean volume:

$$\operatorname{vol}(B(m,r)) = \omega(n)r^n\left(1 - \frac{1}{6(n+2)}\operatorname{scal}(m)r^2 + o(r^2)\right)$$

where $\omega(n)$ is the volume of the unit Euclidean ball in $\mathbb{R}^n$. Unlike the case of Bishop's inequality the condition scal $\geq s$ cannot be integrated in r to yield an inequality for vol $(B(m,r))$. This is shown by easy counterexamples.

It came as a shock when in 1963 Lichnerowicz found a quite strong topological obstruction to scal > 0. The shock can be explained by the following example. Take the Riemannian product $S^2(\delta) \times (M^n, g)$ of any (M^n, g) with the 2-sphere of radius δ. For δ small enough scal$(\delta \otimes g) > 0$ (cf [2], page 86). So that at a superficial glance any topology might be obtained.

The Lichnerowicz result rested on his extension of Bochner's formula to harmonic spinors. In view of formula (7) it is amazing that for any harmonic spinor s one has

$$\frac{1}{2}\Delta(|s|^2) = |\nabla s|^2 + \frac{1}{4}\text{scal}.|s|^2 \ .$$

Amazing because a spinor involves – *a priori* – all the exterior algebra and one would then expect a formula summing over the dimension k the complications of the $F_k(R,.,.)$. In fact at a more conceptual level, J.-P. Bourguignon told me the reason: one has to look for orthogonal-equivariant representation from the space of curvature tensors in the endomorphisms of the exterior algebra. Only the identity is acceptable. So finally the formula above implies immediately that scal > 0 forbids the existence of any harmonic spinors. Of course there is no analogue of the de Rham-Hodge theorem for spinors. But precisely in the very same year, 1963, Atiyah and Singer announced their famous index theorem. As a corollary now scal > 0 implies that the topological invariant called the $\hat{A}$ – genus of (M^n, g) with scal$(g) > 0$ has to vanish. For the philosophy of this, beside what follows, see [20], [40]. That is how one gets the exotic examples of Section 12.

Of course the story does not end there. Let me now recall various recent results in our theme "from scalar curvature to topology".

The first is the converse, or equivalently the characterization of compact M^n admitting at least one Riemannian metric g with scal(g)>0.

After many intermediate (and deep) contributions by Schoen and Yau, Gromov and Lawson, the latest result to mention is that of Stolz; see [44].

The second is that in some cases the condition scal ≥ 0 is extremely sensitive. For example a result of Gromov and Lawson in 1980 asserts that on the n-dimensional torus the condition scal ≥ 0 implies flatness! See also [40].

The third line is to address, in the negative case, the question of the implications of scal $\geq k$ together with some other geometric bounds. This topic is almost completely open now, althrough results are beginning to come out. See [2] and [27] for example.

The last result I wish to quote is that of Schoen and Yau in 1984 (see [2]) on 3-dimensional manifolds. The proof is interesting because of its geometric nature, based on minimal surfaces embedded in the manifold under considerated. This is strongly opposed to Lichnerowicz's method. Their result is as follows: let M^3 be a compact 3-dimensional manifold with non-negative scalar curvature. Then every 2-dimensional homology class can be realized by an embedded oriented surface whose every connected component has non-negative characteristics.

22. Topology from Curvature: Two Topological Invariants Arising from Riemannian Geometry

Here we introduce a basic idea of Gromov's from 1979: use Riemannian geometry to get new invariants for compact manifolds. With a little juggling it still basically belongs to the theme "from curvature to topology".

The first invariant comes quite naturally (think of Section 7). It is defined for a compact M of dimension n as:

$$\mathrm{Min}\|R\|^{n/2}(M) = \mathrm{Inf}\{\|R\|^{n/2}(g) = \int_M |R(g)|^{n/2} d_g m \quad \text{runs through all Riemannian metrics on } M\} .$$

Note that this invariant is indeed dimensionless. In dimension 2 the Gauss-Bonnet formula (Section 4) shows that (except for the absolute value $|K|$ instead of K itself) the invariant is equal to the Euler characteristic of M if K does not change sign. In dimension 4, formula (6) of Section 7 shows that we are in an important domain – that of Einstein manifolds – again linked moreover to the Euler characteristic of M. In fact formula (6) implies immediately that if there exists an Einstein Riemannian metric g on M^4 then it achieves the minimum of $\int_{M^4} \|R(g)\|^2 dm$. For more on this see [17].

Because in dimension 2 the quantity $\int_M k^2 \cdot dm$ is directly linked to the Gauss-Bonnet formula by Schwartz inequality (if one normalizes the volume) and because of its quadratic character, the temptation in dimension n

$$\|R(g)\|^2 = \int_M |R(g)|^2 dm \quad \text{under} \quad \mathrm{vol}(g) = 1 .$$

However a remark of Gromov's kills any hope of getting an interesting result from this consideration: for any $M^n (n \geq 5)$ there exist Riemannian metrics g with $\mathrm{vol}(g) = 1$ and $\int_M |R(g)|^2 dm$ as small as you liked. The construction is wonderfully simple. Start with any Riemannian metric g on M^n and fix a periodic simple geodesic γ in (M^n, g). Now on a tubular neighborhood of γ for g put a flat metric, glue it with what remains and have the radius of the tube larger and larger (for this new flat tube part). The 3-dimensional case depends on what follows.

So we are back to $\mathrm{Min}\|R\|^{n/2}(M)$. To study it directly looked too hard in the late 70's. For that purpose Gromov introduced another invariant:

$$\text{Min Vol}(M) = \text{Inf} \quad \{\text{vol}(g) : g \text{ running through all Riemannian metrics on } M \text{ with } |K| \leq 1\} .$$

Thinking of Section 16 one immediately perceives another domain where it comes in naturally, namely that of compactness results. In other words the condition Min Vol$(M) > 0$ is an insurance against collapsing. This leads one to think that the most important thing is to decide, for a given manifold, when one of the above two invariants is zero and when not. The situation today is as follows.

First it is conjectured that $\text{Min}\|R\|^{n/2} = 0$ is equivalent to Min Vol $= 0$. One direction is trivial because (see Section 2) the fact that R and K are determined one by the other implies

$$\text{Min}\|R\|^{n/2} \leq c(n)\text{Min Vol } (c(n) \text{ depending only on the dimension } n) .$$

The other way around would require studying Riemannian manifolds (M^n, g) with $\|R(g)\|^{n/2}$ going to zero, a story making progress but still not completely clear (see next section).

Secondly, for the invariant Min Vol there are many examples of manifolds with Min Vol > 0 and many with Min Vol $= 0$ (see [31] and [36]) but there is no reasonable conjecture to date on the real line of separation. There is not a single odd dimensional simply connected manifold with Min Vol > 0 known today!

Thirdly, there is a conjecture that the value 0 is isolated in the set $\{\text{Min Vol}(M^n) : M^n$ running through all compact manifolds of dimension $n\}$. There would exist $\varepsilon(n) > 0$ with Min Vol $(M^n) < \varepsilon(n)$ implying Min Vol$(M^n) = 0$. This is now proved when $n = 3$ as a corollary of [36].

To give the reader an idea of the subtlety of the invariant we mention that its exact value on a given manifold is unknown for any manifold of dimension ≥ 5. E.g. one would expect Min Vol(S^n),

$\operatorname{Min}\|R(g)\|^{n/2}(S^n)$ to be equal to the value obtained by the canonical metric on S^n. And the same goes for standard spaces such as space forms, generalized space forms, etc. For more on these invariants the reader is referred to [17], [31], [36], [45].

23. Harmonic Analysis on the Set of Riemannian Manifolds: Integral Norms on Curvature

For ordinary numerical functions there are famous and basic relations between the sup norm and the L^p-norms of the function, its gradient and its derivatives. They are classically called Sobolev's inequalities. They are basic in many respects. For example, when one knows that a function obeys inequalities of another type, e.g. coming from the fact that they are solutions of partial differential equations, then one can deduce finally the sup norm (L^∞) from only one L^p norm by a bootstrap method.

So it is natural in "from curvature to topology" still to expect results when one replaces conditions like $K \geq k$, etc. by integral conditions like

$$\int_M |R(g)|^p dm \leq a$$

where $|R|$ means the full norm of the curvature tensor namely

$$|R|(m) = \sqrt{\sum_{i,j,k,h} R^2_{ijkh}(m)}$$

(where R_{ijkh} denote the components of R in any orthonormal basis) and analogous notations for Ricci curvature. Remember (Section 2) that knowledge of K and of R are basically equivalent when one is not seeking a precise value.

To our knowledge, results in this direction started coming out only very recently. Note also that the case of the $L^{n/2}$ norm naturally comes in the problem we address in the preceding section. So the big idea would be to extend the technique mentioned above to Riemannian metrics. The first of these results appears in the late 1980's. They obviously needed the following intermediate-type result. A typical inequality for a function f on a given Riemannian manifold of dimension n is $\|f\|_p \leq A\|df\|_q + B\|f\|_p$ with $\frac{1}{p} + \frac{1}{q} = \frac{1}{n}$ where df is the differential of f. But as soon as one wants to work in the realm of *all* Riemannian manifolds the problem lies in estimating A and B as functions of the Riemannian invariants of (M^n, g). Just as A and B for functions in Euclidean space are obtained through the classical isoperimetric inequality, so in a Riemannian manifold (M^n, g) one will need "isoperimetric inequality" for Riemannian manifolds, where the constants in various such isoperimetric inequalities should be estimated with the weakest possible Riemannian invariants, at most $\sup|K|$, inf Inj, at best inf Ric, inf vol.

We cannot give details here on the interplay between recent results on this topic. Let us just mention the names of T. Aubin, J. Cheeger, C. Croke, S. Gallot, M. Gromov, P. Li, S. T. Yau.

An optimal version for the isoperimetric inequality involving only inf Ric and diameter was obtained in 1985 by Bérard-Besson-Gallot. Recently Gallot in [39] succeeded in getting even isoperimetric inequalities based only on the L^p norm $\|\mathrm{Ric}\|^p$ but the condition $p > \frac{n}{2}$ cannot be replaced by $\frac{n}{2}$ as counterexamples show.

Then – as we have said above – results bearing this time directly on the space of Riemannian manifolds started to appear in the late 1980's. The topic being still in a state of evolution we will just mention a few of them. In strong opposition to preceding results on the implications from curvature conditions to topology, where techniques were mostly of a geometrical nature, the tools here are analytical. One of the major tools is the deformation under the Ricci flow: one uses the fact that the Ricci curvature belongs to the same function as the space of Riemannian metrics, namely symmetric bilinear differential forms.

This flow is a time dependent family of Riemannian metrics $g(t)$ such that

$$\frac{\partial(g(t))}{\partial t} = -2k \operatorname{Ric}(g(t))$$

where k has to be adjusted nicely. Hamilton, Bemelmans, Min-Oo and Ruh first discovered in 1984 ([50]), that this flow has the effect of smoothing the entire curvature tensor. So it has since become a basic tool for people working with integral norms of the curvature and it can be considered as a geometrico-analytical pendant of the purely geometric triangle comparison theorem. It was first introduced by Hamilton in 1982 to study 3-manifolds with positive Ricci curvature: [52]. See also [53].

Today results are of two types. Typical of the first type are Gao's pinching theorems of 1990 ([46], [47]) for manifolds lying in some $\{\operatorname{Ric} \geq \bullet, \operatorname{Inj} \geq \bullet, \|R\|^{n/2} \leq \bullet\}$ where the pinching is in L^2 – norm for the deviation of the curvature tensor from the positive constant sectional case. For the zero and the negative cases, Gao has to add diameter $\leq \bullet$. The flat case is a step toward the open question quoted in the preceding section concerning $\operatorname{Min}(\|R\|^{n/2})$. Yang's result in [51] is another such step. The general scheme is always to deform the given metric under a suitable curvature-driven flow into a metric for which the conclusion is already known..

Of the second type there are the following results due to Gallot. They give universal bounds for all real Betti numbers of a Riemannian manifold in the sole functions $\|\operatorname{Ric}\|^{p/2}$ and diameter as soon as one $p > n$ is given. The counterexamples of Sha-Yau in Section 20 show that $p = \frac{n}{2}$ does not work. Contrary to Gao's results but similar to Gromov's results in Section 15, it is to be noted that no lower bound on injectivity radius is necessary. But here only the non torsion part of the homology is accessible because Gallot essentially exploits Bochner-Lichnerowicz formulas.

Bibliography

[1] W. Ballman, M. Gromov, V. Schroeder, *Manifolds of non-positive curvature*, Progress in Math. vol. 61, Birkhaüser 1985.

[2] M. Gromov, *Sign and geometric meaning of curvature*, preprint IHES 1990.

[3] T. Sakai, *Comparison and finiteness theorems in Riemannian geometry, in Geometry of Geodesics Related Topics*, Advanced Studies in pure Math. 3, North Holland 1984.

[4] M. Berger, B. Gostiaux, *Differential geometry*, Springer 1988.

[5] M. Spivak, *Differential geometry II*, Publish or Perish.

[6] J. Wolf, *Spaces of constant curvature*, Publish or Perish.

[7] H. Hopf, *Differentialgeometrie und topologische Gestalt*, Jahresbericht d. DMV 41 (1932) 209-229.

[8] H. Hopf, *Über die Curvatura integra geschossener Hyperflächen*, Math. Annalen 95 (1925) 340-367.

[9] H. Hopf, *Die Curvatura integra Clifford-Klenischer Raumformen*, Nachr. Ges. d. Wiss, Göttingen, Math. Phys. Klasse 1925.

[10] J. Cheeger, W. Müller, R. Schrader, *On the curvature of piecewise flat spaces*, Comm. Math. Phys. 92 (1984) 405-454.

[11] J. Hadamard, *Les surface à courbures opposées et leurs lignes géodésiques*, J. Math. pures et appl. 4 (1898) 27-73.

[12] A. Preissmann, *Quelques propriétés des espaces de Riemann*, Comm. Math. Helv. 15 (1942-43) 175-216.

[13] J. Hadamard, *Sur certaines propriétés des trajectoires en dynamique*, J. Math. pures et appl. 3 (1897) 331-387.

[14] M. Spivak, *Differential geometry V*, Publish or Perish.

[15] A. Gray, *Tubes*, Addison-Wesley 1990.

[16] S.S. Chern, *On the curvature and characteristic classes of a Riemannian manifold*, Abh. Math. Sem. Hamburg 20 (1956) 117-126.

[17] A. Besse, *Einstein manifolds*, Springer 1987.

[18] J.-P. Bourguignon, A. Polombo, *Intégrands des nombres*

caractéristiques et courbure: rien ne va plus dès la dimension 6, J. Diff. Geometry 16 (1981) 537-550.

[19] N. Berline, E. Getzler, M. Vergne, *Heat Kernels and Dirac operators*, Springer 1991.

[20] B. Lawson, M.L. Michelsohn, *Spin geometry*, Princeton University Press 1989.

[21] J. Cheeger, D. Gromoll, *On the injectivity radius of 1/4-pinched manifolds*, J. Diff. Geometry 15 (1980) 437-442.

[22] H. Karcher, *Riemannian comparison constructions*, Global Differential Geometry, Studies in Math. 27, Mathematical Ass. of America 1989.

[23] J. Eschenburg, *Local convexity and non-negative curvature*, Inv. Math. 84 (1986) 507-522.

[24] M. Micallef, J. Moore, *Minimal two-spheres and the topology of manifolds with positive curvature on totally isotropic two-planes*, Annals of Math. 127 (1988) 199-227.

[25] M. Gromov, *Stability and pinching*, preprint IHES 1990.

[26] M. Gromov, P. Pansu, *Rigidity of lattices: an introduction*, Springer Lecture Notes in Math.

[27] M. Gromov, *Foliated plateau problem*, Geometric and Fucntional Analysis 1 (1990).

[28] L. Hernandez, *Kähler manifolds and 1/4-pinching*, Duke Math. J.

[29] J.-H. Eschenburg, *Compact spaces of positive curvature.*

[30] E. Heintze, H. Karcher, *A general comparison theorem with applications to volume estimates for submanifolds*, Ann. Scient. E.N.S. 11 (1978) 451-470.

[31] M. Gromov, *Volume and bounded cohomology*, Publications Mathématiques IHES 56 (1983) 213-307.

[32] M. Gromov, J. Lafontaine, P. Pansu, *Structures métriques pour les variétés riemanniennes*, Cedic-Nathan 1981.

[33] S. Peters, *Convergence of Riemannian manifolds*, Comp. Math. 62 (1987) 3-16.

[34] T. Yamaguchi, *On the number of diffeomorphism classes in a certain class of Riemannian manifolds*, Nagoya Math. J. 97 (1985) 173-192.

[35] I. Nikolaev, *Bounded curvature closure of the set of compact Riemannian manifolds*, Bulletin A.M.S. 24 (1991) 171-178.

[36] J. Cheeger, M. Gromov, *Collapsing Riemannian manifolds while keeping their curvature bounded I; II*, J. Diff. Geometry 23 (1986) 309-346; 32 (1990) 269-298.

[37] K. Grove, K. Shiohama, *A generalized sphere theorem*, Ann. Math. 106 (1977) 201-211.

[38] M. Gromov, *Curvature, diameter and Betti numbers*, Comm. Math. Helv. 56 (1981) 179-195.

[39] S. Gallot, *Isoperimetric inequalities based on integral norms of Ricci curvature*, Astérisque 157-158 (1988) 191-216.

[40] M. Gromov, B. Lawson, *Positive scalar curvature and the Dirac operator on complete Riemannian manifolds*, Publications Mathématiques IHES 58 (1983) 295-408.

[41] K. Grove, P. Petersen V, *Bounding homotopy types by geometry*, Annals of Math. 128 (1988) 195-206.

[42] K. Grove, P. Petersen V, J-Y Wu, *Geometric finiteness theorems via controlled topology*, Invent. Math. 99 (1990) 205-213.

[43] K. Fukaya, *Collapsing Riemannian manifolds to ones of lower dimension*, to appear.

[44] S. Stolz, *Simply connected manifolds of positive scalar curvature*, Bulletin A.M.S. vol. 23 (1990) 427-432.

[45] J. Cheeger, K. Fukaya, M. Gromov, to appear.

[46] L. Gao, *Convergence of Riemannian manifolds; Ricci and $L^{n/2}$-curvature pinching*, J. Diff. Geometry 32 (1990) 349-382.

[47] L. Gao, *$L^{n/2}$-curvature pinching*, J. Diff. Geometry 32 (1990) 713-774.

[48] S. Peters, *Cheeger's finiteness theorem for diffeomorphism classes of Riemannian manifolds*, J. Reine Angew. Math. 394 (1984) 77-

82.

[49] J. Cheeger, D. Ebin, *Comparison theorems in Riemannian geometry*, North Holland 1975.

[50] J. Bemelmans, M. Min-Oo, E. Ruh, *Smoothing Riemannian metrics*, Math. Z. 188 (1984) 69-74.

[51] D. Yang, *Riemannian manifolds with small integral norm of curvature*, to appear.

[52] R. Hamilton, *3-manifolds with positive Ricci curvature*, J. Diff. Geometry 12 (1982) 255-306.

[53] C. Margerin, to appear.

[54] G. Besson, G. Courtois, S. Gallot, *Volume et entropie nominale des espaces localement symétriques*, Invent. Math. 103 (1991) 417-445.

[55] Demailly, *Analytic and algebraic geometry*, to appear.

[56] F. Hirzebruch, *Topological methods in algebraic geometry*, 3rd ed. Springer-Verlag 1966.

[57] K. Grove, *Metric differential geometry*, in Differential Geometry, ed. Hansen, Springer Lecture Notes in Mathematics n° 1263, 1985.

[58] P. Dombrowski, *Differentialgeometrie*, in Dokumente zur Geschichte der Mathematik, vol. 6, Vieweg 1990.

[59] H. Wu, *The Bochner technique in differential geometry*, Math. Reports, London 1987.

[60] P. Bérard, *From vanishing theorems to estimating theorems; the Bochner technique revisited*, Bulletin A.M.S. vol. 19 (1988) 371-406.

[61] J.-P. Bourguignon, *The "magic" Weitzenböck formulas*, in Variational methods, Progress in nonlinear differential equations and their application vol. 4, Birkhaüser 1990.

[62] L.Z. Gao, S.T. Yau, *The existence of negatively Ricci curved metrics on three manifolds*, Inventiones Math. 85 (1986) 75-90.

[63] J.-P. Sha, D.-G. Yang, *Examples of manifolds of positive Ricci curvature*, J. Differential Geometry 29 (1989) 95-103.

[64] R. Greene, H. Wu, *Lipschitz convergence of Riemannian manifolds*, Pacific J. of Math. 131 (1988) 119-141.

[65] J. Cheeger, *Finiteness theorems for Riemannian manifolds*, Amer. J. Math. 92 (1970) 61-74.

[66] K. Fukaya, *Hausdorff convergence of Riemannian manifolds and its applications*, in Recent topics in Differential and Analytic Geometry, ed. T. Ochiai, Tokyo 1990.

[67] J. Cheeger, *Critical points of distance functions and applications to geometry*, to appear.

[68] P. Dombrowski, *150 years after Gauss*, Astérisque n° 62 Société Mathématique de France 1979.

[69] R. Osserman, *The isoperimetric inequality*, Bulletin A.M.S. 84 (1978) 1182-1238.

[70] S. Gallot, *Inégalités isopérimétriques et analytiques sur les variétés riemanniennes*, in On the Geometry of Riemannian manifolds, Astérisque 163-164 Société Mathématique de France 1988.

[71] P. Bérard, M. Berger, *Le spectre d'une variété riemannienne*, in Spectra of Riemannian manifolds, ed. Berger, Murakami, Ochiai, Kaigai 1983.

[72] H. Gillet, C. Soulé, *An arithmetic Riemann-Roch*, To appear.

[73] M. Gromov, *Filling Riemannian manifolds*, J. Differential Geometry 18 (1983) 1-147.

[74] Loday, *Cyclic homology*, Springer 1992.

[75] C. Lanczos, *A remarkable property of the Riemann-Christoffel tensor in four dimensions*, Ann. Math. 39 (1938) 842-850.

[76] F. Farell, L. Jones, *Negatively curved manifolds with exotic smooth structures*, J. Ann. Math. Soc. 2 (1989) 1-21.

[77] D. Brittain, *A diameter pinching theorem for positive Ricci curvature.*

[78] A. Katsuda, *Gromov's convergence theorem and its applications*, Nagoya Math. J. 100 (1985) 11-48 and corrections Nagoya Math. J. 114 (1989) 173-174.

[79] Z. Shen, *On complete manifolds of nonnegative kth-Ricci curvature*, To appear.

[80] Y. Otsu, K. Shiohama, T. Yamaguchi, *A new version of the differentiable sphere theorem*, Inventiones Math. 98 (1989) 219-228.

[81] F. Kamber, P. Tondeur, *Foliated bundle of characteristic classes*, Springer Lectures Notes in Mathematics 493, 1975.

[82] J. Dupont, F. Kamber, *Cheeger-Chern-Simons classes of transversally symmetric foliations: dependence relations and eta-invariants*, Preprint Aarhus U. 1989/1990.

[83] L.Bérard Bergery, *Les variétés riemanniennes homogènes simplement connexes de dimension impaire à courbure strictement positive*, J. Math. pures et appl. 55 (1976) 47-68.

[84] M. Gromov, R. Schoen, *Harmonic maps into singular spaces and p-adic superrigidity for lattices in groups of rank one*, Publications Math. de l'IHES 1992.

[85] U. Abresch, D. Gromoll, *On complete manifolds with nonegative Ricci curvature*, J. of the Amer.Math. Soc. 3 (1990) 355-374.

On Chern and Youth

Blaine Lawson
Department of Mathematics
SUNY

It is difficult for me to find where to begin discussing Chern who has deeply affected my life in many ways. His direct influence began at the end of my graduate studies. I thought it would be interesting to talk about this since it illustrates some of Chern's marvelous personal qualities.

My first meeting with Chern was for me a pretty upsetting experience. I was still a graduate student at Stanford at what was, I hoped, the end of my graduate career. It was Spring and Chern had come to the Joint Stanford Berkeley Colloquium. During the tea preceeding the talk, I overheard some fragments of conversation emanating from the circle around Chern which had a strikingly familiar ring. I moved closer and began to listen attentively. To my utter astonishment, Chern was discussing in minute detail several results from my thesis. I was astonished because to date these results had never passed outside my own notebooks and discussions with my advisor. My heart began to sink as I realized that a group of geometers at Berkeley (Chern, do-Carmo, and Kobayashi) had carried forward some of the recent results of Jim Simons exactly as I had. With words I don't remember but emotions that are still vivid, I mentioned that I also proved some similar theorems. I will never forget the grace and charm with which Chern received me. What I first assumed was the total loss of a significant part of my thesis was changed in a matter of weeks into an offer of an appointment at Berkeley. Furthermore, from that moment on, Chern never talked about these results without mentioning my name.

From the moment I arrived at Berkeley I was welcomed as a colleague, as an equal among the people in geometry. This was a strange but a very heady experience. It felt ridiculous for a kid with the ink still wet on his thesis to be accepted into a group of such towering intellectual achievement. It also felt very good.

This atmosphere of collegiality was set by Chern whose openness and respect for young people was exceptional. It was manifest in many ways. He was truly interested in what we had to say. He listened; he questioned; he even asked for opinions and advice. His attitude extended to all the young people in the field. It is my belief that this excitement of feeling part of the real world of mathematics, which Chern so quietly but so effectively engendered in young people, greatly contributed to the flowering of geometry in Berkeley at that time.

Chern's belief in young people and his encouragement of them had, in fact, a lot to do with the spectacular growth of geometry in the second half of this century. (This is a factor quite apart from the influence of his work, whose central place in twentieth century mathematics is clear to anyone reading this.) It is not easy to find a geometer who was not for some period of time either a student or a post-doctoral fellow in the orbit of Chern.

This belief of Chern in young people was quite real and complete. It would otherwise never have been so effective. It extended equally to women and men. It gave the geometry group at Berkeley a special atmosphere, a wonderful one in which to live and work.

It was Chern who welcomed me into the family of research mathematicians, and I, like a new-born goseling, implicitly believed that his very special world simply reflected the general reality. For that gift I will always be deeply grateful.

Remarks delivered at Chern 79th Birthday Celebration

Jeff Cheeger
Courant Institute
New York

Its an honor to have been asked to say a few words at this birthday celebration for Professor Chern. If much of what I say seems to have a purely personal interest, I ask your understanding. My intention is to illustrate a general principle by examining a not atypical special case, naturally, the one with which I'm most familiar.

Chern's work in geometry has, of course, been a great source of inspiration for my own, with Simons for example, with Gromov and with Müller and Schrader. Indeed, who among us could not make a similar claim.

But in actuality, our contact at the human level has been the more decisive influence, and it would not surprise me if the same holds true for many others as well.

I met Chern in 1967, shortly after having received my first job offer, an instructorship at Berkeley. After we were introduced, Chern asked in an understated manner which I later realized was quite characteristic, whether I thought there was "any chance" that I "might" accept their offer? What came through was that here, for the first time, I was meeting the great Chern, and that *he* was, going out of his way to make me feel that he personally wanted me to come to Berkeley.

Shortly after I arrived, I went to Chern's office to say hello. I remember that he greeted me warmly and that we chatted for some time, as much on nonmathematical subjects as on mathematical ones. Again I was struck by the fact that he seemed genuinely interested in my ideas and opinions (however ill formed).

Towards the end of our conversation he said quite directly, "You should talk to Gromoll about the work he's been doing with Meyer on complete open manifolds of positive curvature" (I recall becoming momentarily confused and asking if he didn't mean negative curvature.)

Armed with Chern's suggestion I felt bold enough to introduce myself to Gromoll and in short order, we entered a collaboration which was to be of the utmost importance for my scientific development.

Maybe I would eventually have talked to Gromoll anyway. But in my mind I've always felt that quite possibly, maybe I wouldn't. And then a lot of things would have turned out very differently for me. In any case, I like to think that Chern wasn't taking any chances.

The year I spent in Berkeley was in many ways and for many different reasons, one of the most exciting of my life. But the bottom line was that it was scientifically productive, at a stage when my career was still quite fragile, and this I owe in large measure to Chern's influence and guidance.

Over the years, Chern has continued to express an interest in my work and to provide encouragement. Coming from him, it's meant a lot.

One of the real privileges of our profession is the opportunity to come in contact with some of the most brilliant minds of our time. But occasionally these individuals seem to delight in scaring half to death, those less gifted than themselves.

At the opposite end of the spectrum are men like Chern, who by his council, his teaching and his own magnificent example, inspires others to reach greater heights than they themselves might have believed possible.

To be a great mathematician is not the same as to be a great man. Chern is a great mathematician *and* a great man.

I just want to wish him a happy birthday and to say "Thanks for everything".

Some Thoughts About S.S. Chern

Alan Weinstein
Department of Mathematics
University of California, Berkeley

On the mathematical side, I learned from Chern, to my astonishment, how much one could accomplish by writing down some obvious equations in forms and frames, applying "*d*" to them, substituting, and then doing it all over again. Why can't we all make a living (and much more!) this way? It was some time before I realized that Chern knew all along what a principal bundle was.

Perhaps the most important lesson I learned from Chern was how to be a teacher. Hardly a month goes by when I fail to recall how much time he seemed to have to talk when I was a graduate student, even before I became "his" student. Here he was, one of the new "stars" in the Berkeley department, with a pile of mail and other papers on his desk as large as any I have ever had on my own (or in my electronic mail box), and yet he never seemed to be so much in a hurry to get on with his work that he didn't have time to listen to all I had to say. He prolonged our conversations by talking about his own work and even asking me for my advice.

To approach Chern's accomplishments as a mathematician is something I don't even dream of, but the generous way in which he has encouraged and supported so many geometers provides an ideal which appears at first more approachable. Like the magic with differential forms, though, Chern's human spirit turns out to be uniquely his own.

Chern likes to refer to Chinese history and philosophy when explaining his attitude toward life. You can recognize him immediately in the following quotation from Lao Tzu, in the beautiful translation by Stephen Mitchell.

The Master does his job
and then stops.
He understands that the universe
is forever out of control,
and that trying to dominate events
goes against the current of the Tao.
Because he believes in himself,
he doesn't try to convince others.
Because he is content with himself,
he doesn't need others' approval.
Because he accepts himself,
the whole world accepts him.

Lao Tsu
Tao Te Ching
Translation by Stephen Mitchell
@ 1988

S.S. Chern:
Some Mathematical and Personal Reminiscences

by Robert E. Greene
Department of Mathematics
UCLA

In the Nineteenth Century, comparatively few people were engaged in mathematical research, and our vision of the major developments of that century as having emanated primarily from a handful of heroic figures is reasonably accurate. In our own century, mathematical research has expanded to the point that the subject is more realistically regarded as a continuous, communal activity than as a sequence of steps each partaking of individual heroic effort. The age of the giants might well seem to have ended back at the beginning of this century. But there have been some exceptions, even in our own times. Occasionally, if rarely, a single figure has emerged from the crowd of active mathematicians to become by consensus the acknowledged leader of some broad area of mathematics over decades. Such is the case with S.S. Chern and differential geometry. Mathematical leadership exists not in the trappings of official authority, but in the minds and mathematical hearts of one's colleagues. It is a kind of composite of the respect and appreciation of many individuals. I am both pleased and honored to express my own appreciation, describing here how my mathematical life has been enriched by the work and the personal inspiration of Professor Chern[2]. At the risk of talking a bit about myself, I can best express such an appreciation, I think, by presenting my personal experiences and discussing Chern's influence on my own work, rather than writing generalities.

I first encountered Chern in 1965, when I began my graduate study at the University of California, Berkeley. At this point, I was only very partially aware of the specifics of Chern's mathematical accomplishments, but naturally his reputation among the graduate students approached legendary proportions. Chern's lectures on intrinsic invariants and in particular on his intrinsic proof of the generalized Gauss-Bonnet formula were a startling tour de force. Complex calculations unfolded on the blackboard without notes and without apparent effort, and also without errors. (Later on, when I was beginning research under the direction of H. Wu, I asked him whether this kind of almost supernatural computational skill was a prerequisite for a successful career in geometry, and I was much relieved to be assured that I did not have to answer to such a standard.) But Chern's awesome lectures were delivered with an obvious, sincere modesty. At this time for example, Chern invariably referred to Chern classes as "the so-called Chern classes", and generally seemed to consider his own accomplishments as not worthy of exceptional notice. Even so, one day one of the students could not resist asking him about the blackboard displays we had been watching lecture after lecture, asking in effect if these were prepared inch-by-inch in advance as a sort of theater piece. Apparently baffled by such a concept, Chern said "Nowadays when I have something really

[2]Mathematicians are an informal bunch, not given to refering to each other by titles. But everyone in geometry refers to Professor Chern just so, as "Professor Chern", even in his absence. In this article, I shall call him just Chern for brevity, but if you and I were talking, or any geometers anywhere were, "Professor Chern" it would be. This honor, to be thus referred to by title in everyday conversation, is unique to Professor Chern, in my observations.

complicated to do, I do use a little paper. But when I was younger, I never had to write down anything....., " after which we were even more awestruck than before.

After a brief flirtation with topology, I decided to take up geometry. At this same time, I began to develop the mathematical rapport with H. Wu which led to our long-term joint work (over the intervening twenty years!); and I pursued my thesis research under his direction. We were both much under Chern's influence and benefitted from his support. And Wu's and my first joint project, on the rigidity properties of punctured nonnegative curvature surfaces, was directly related to some of Chern's earlier work: The crucial ingredient in our rigidity proof was establishing the convexity of the embedded surface. In the unpunctured case, this convexity had been the subject of a paper of Chern and R. Lashof; and, using the convexity they proved, K. Voss and independently R. Sacksdeder had proved rigidity in the unpunctured case. Wu's and my second paper on this subject appeared in the volume of the Journal of Differential Geometry dedicated to Chern on his sixtieth birthday. It was a source of great satisfaction to both Wu and me to participate in this occasion for honoring Chern, the first of several such satisfactions as it turned out.

In the early 1970's, Wu and I began what turned out to be a longterm project on the relationship between curvature and function theory of complex manifolds. Learning about complex manifolds then began most effectively by reading Chern's lecture notes on the subject, which I had studied as a graduate student. Then, and now — — —— Chern's Complex Manifolds without Potential Theory has reappeared in a second edition, published by Springer in 1979, to guide new generations of students into the subject. The natural setting in which to begin complex function theory of noncompact Kahler manifolds is the topologically trivial case, in which characteristic classes as such play no role. But the ideas of Hermitian curvature of bundles and related matters that Chern had used to describe Chern classes geometrically turn out to be the appropriate setting in the function-theoretic case, too. In particular, the association between bundle curvature and the L^2 d-bar method together with the relationship between Kahler manifold geometry and bundle curvature provided a basic connection between geometry and function theory. On a more personal level, Chern's advice and encouragement were a great help to us. Even though we were working on a subject apparently quite far in its technical details from Chern's own direct interests, he invariably had insights to offer and suggestions always worth following. This breadth of mathematical vision, of insight into the general pattern that a subject must follow, it surely one of the characteristics of true mathematical leadership.

In the late 1970's, I began a project with Steven Krantz of applying geometric methods to the function theory of domains in Euclidean spaces. Here again, there was a strong interaction with Chern's ideas. Our goal was to understand biholomorphic mappings from the viewpoint of the intrinsic (interior) geometry of the Bergman metric. Chern's invariant theory, extending ideas of Poincare and Tanaka, and developed in another way also by Moser, analyzes similar questions from the geometry of the extensions of the mappings to the boundary of the domain. These viewpoints are connected by Fefferman's result that the asymptotic behavior of the Bergman metric is determined by the Chern-Moser invariants of the boundary. Moreover Fefferman's theorem that biholomorphic maps of smooth strongly pseudoconvex domains extend smoothly to the boundary insures that the boundary invariant theory applies. It is worth noting that Chern (and Moser) with an almost uncanny sense of timing reactivated and extended the historic Poincare hypersurface invariant theory just a year or so before Fefferman proved the quoted result that shows that the hypersurface invariant theory actually applies to biholomorphic mappings. Since the group of biholomorphic mappings (of a smooth strictly pseudoconvex domain) thus acts on the boundary, the Chern-Moser boundary invariant theory enables one to study the automorphism group via the geometry, in the Chern-Moser sense, of the boundary. Krantz's and my hope, which was eventually realized, was to develop a similar analysis based on the metric geometry of the interior of the domain, relative to the Bergman metric geometry, which is invariant under biholomorphic mappings. In this picture, the Chern-Moser boundary geometry is, in a suitable sense, a compactification of the Bergman interior metric geometry. In effect, Krantz and I were working to develop a de-compactification, or interiorization, of the Chern-Moser theory.

What I have been recounting here – how most of my mathematical life has consisted of exploration of ideas that Chern either originated or contributed largely to – this is, in fact, the common experience of my generation of geometers. And of course there are other directions of Chern's work that have not figured so much in my own work, but have been of signal significance to others: minimal submanifolds, differential systems, Chern-Simons invariants, and so on. Naturally, the geometry community has been inclined to express its indebtedness with various events in his honor, including his retirement celebration at Berkeley in 1979, and a seventieth birthday celebration at UCLA in 1981, which I had the pleasure of organizing, with S.Y. Cheng. Although these events had limited funding and had to be attended largely on a volunteer basis, they still attracted large numbers of geometers and had a true spirit of a gathering of the clan to honor our leader. And the 1991 American Mathematical Society Summer Institute on Differential Geometry at UCLA took on something of the same spirit: Chern gave the opening address; the organizing committee decided to dedicate the published proceedings to him; a special one-day event was organized in his explicit honor, this latter being the origin of the present volume.

This kind of respect and recognition cannot accrue to someone from technical mathematical accomplishments alone, however impressive those might be. A strong component of moral character must be involved. Chern has always used his influence "with firmness in the right," and he has earned the moral respect of his colleagues over decades. In addition, he has been an inspiration not only in the excellence of his work and grandness of his geometric vision but also in the steadfastness of his dedication.

Mathematics is kind to youth but often takes its toll with the passing years. The feeling that one's work has reached its peak in quality or that one's reputation has stabilized, the flagging of the competitive spirit, a shortage of the energy required to keep up with the rapid development of the subject, the ennui that can arise from study of a subject remote from ordinary life, the sense that younger people really want to have the scene to themselves – – –– there are many reasons for the familiar phenomenon that many mathematicians bring their research careers to a close somewhere in middle age. But Chern has overcome all the possible problems, it seems, through sheer love of the subject on the merits of its intellectual glory. In recent years, he has, with characteristic modesty, taken to describing what he works on as "old man's problems". But in reality, his remains an active, vital, and profound mathematical mind. His life and work show that mathematics, too, can have the special quality more usually found in art and music, that the works of one's later years have a uniquely valuable character, reflecting the accumulated wisdom of a lifetime. In a world obsessed with youth and the accomplishments of youth, Chern's later mathematical life has been a continuing inspiration towards mathematics as life-long, life-filling, and life-enhancing. This inspiration is a gift to us of extraordinary value, one we much needed and for which we should be deeply grateful.

My teacher Professor S.S. Chern

Shiu-Yuen Cheng
Department of Mathematics
UCLA

I am a student of Chern and most of my mathematical friends are either students or grandstudents of Chern. Stories about Chern frequently appear in our conversations. Over all these years, I have been able to collect many stories about Chern. However, when I started writing this article. I found it difficult to link all these stories together. After some thought. I have decided to do it chronologically. I find that my earlier impressions are much more vivid in my memory, so I shall write mostly about those.

I entered United College of the Chinese University of Hong Kong in the Fall of 1966, majoring in mathematics (Yau entered Chung Chi College at the same time). I made this choice because I had a good mathematics teacher in high school and I really enjoyed Plane Geometry and Algebra. Also, there weren't many other choices; at United College, Mathematics, Physics and Chemistry were the only majors in science. At that time the three colleges of the Chinese University of Hong Kong were at different locations. Chung Chi was located at the present site of the Chinese University and had the most beautiful campus. The campus of the United College was the worst; our next dear neighbour was a psychiatric institute and we shared the same basketball court. Naturally, the library had few mathematics books. I hadn't the faintest idea about the field of mathematics and the leading mathematicians. We just assumed that the textbook authors were the major figures. Chern did not write any textbooks, and hence I hadn't heard of him. The first time I heard about Chern was in 1967. I found an article by Chern in the second issue of Ming Pao Magazine. Its title was "Forty years of mathematics" and was a short autobiography. The issue had been published a year earlier and since it was a new magazine I had never seen it before. Chern's article was fascinating. It was in chronological order beginning from his childhood. One of the most interesting episodes during Chern's college years was when Chern's career as an experimental scientist ended in his freshman year when he tried to cool down the newly blown glass pipes with tap water. His classmate told me that one could always spot Chern during exercise drills in physical education classes. He was the one not synchronized with the others. I think Chern can be quite handy if he chooses to. He once advised us to appear incompetent doing household chores. Then we would never be asked to perform these jobs.

The article also opened my eyes in many ways about mathematics. It was when I first heard of differential geometry. Much of the terminology such as fiber bundles and Chern characteristic classes was new to me. I had no idea of the meanings of the phrases but the Chinese translations were fascinating. Reading the article I also found that Chern might have known my father (my father passed away when I was seven) since they attended Nankai University around the same time. To be able to relate to one of the world's greatest mathematician, even through this indirect way, still was very exciting. It turns out that Chern did know my father, though my father was several years older than him. However, I could not imagine that I would have a chance to study under Chern at the University of California, Berkeley. Chinese University was quite new then, so it hadn't yet a chance to establish a good track record with its students. This was to be changed completely by the success of Yau.

Chern visited Hong Kong in the Summer of 1969 and gave a seminar. This was very rare; during my four years in college, only a handful of mathematicians passed by Hong Kong to give seminars. For these rare occasions, many of us undergraduates tried to attend though we didn't even understand the titles. Chern talked about the Gauss-Bonnet theorem and characteristic classes. All I understood was that the Gauss-Bonnet theorem is a generalization of the well-known theorem that the sum of a plane triangle's interior angles is 180. This was very fascinating; I really wanted to see how a theorem in high school geometry could be connected with deep results in modern mathematics. I asked Yau about it since he had read something on classical differential geometry. I remember that I did not get a satisfactory answer. After the talk, I was too shy to talk to Chern, but I saw Chern was talking to Yau, who was about to enter UC Berkeley for graduate study.

In 1970, Ming Pao published an interview with Chern. Chern told the interviewer that he was working on minimal submanifolds and even briefly described what the subject was about. I was attracted by the description of Chern's house and its fantastic view of the San Francisco bay. Later, I found that the scenery was better than the description in the article. Also, Yau wrote to me about Berkeley and Chern. Since Chern was a Miller professor (a nonteaching position) in 1969, Yau didn't have many chances to see him. Using a Chinese proverb, Yau wrote that Chern's position in Differential Geometry was like that of the Tai Mountain (a famous mountain in China) and the Big Dipper. In early 1970, I heard from some students that Chern wrote a letter to the Chinese University. The letter stated that Yau had done something significant and that Yau would become a leading mathematician. Yau's success helped a lot of students from Hong Kong to get admissions to U.S. Universities. I was one of the beneficiaries; a classmate and I were admitted to Berkeley with fellowships.

I arrived in Berkeley in mid August. Chern was away at the International Congress in Nice to give his plenary address. In September, he came back and Yau took me to see him. Chern told us about the Congress, and I said very little during the whole conversation since my Mandarin was very poor. Afterwards we walked with Chern from Campbell Hall to the parking lot on Hearst Avenue. Chern told me that he usually parked his car far away and this walk was his exercise. That was not in fact a long walk at all. I have always been amazed by Chern's physical ability. He has no diet restrictions, he doesn't need to take exercise, yet he works hard and travels a lot. I still remember one incident in the Summer of 1985 when I attended the Summer School at Nankai organized by Chern. Nankai sent a van to pick up Chern from the Beijing airport. I also went. It was past midnight when the plane arrived and the ride back from Beijing to Tianjin was more than three hours. However, even after many hours of international flight and time zone changes, Chern was still very energetic and talking all the way back to Tianjin.

One of the most memorable event during my first quarter at Berkeley was Chern's annual Christmas party for all the Chinese students. Mrs. Chern cooked food for about thirty people. This was the first time I got to see Chern's famous house. It was a very relaxed atmosphere there. We talked casually with Prof. Chern and Mrs. Chern. Chern and Yau played a game of Chinese chess. I later asked Yau about the result. He claimed that he won but I had no way to verify this. As far as I know Yau hasn't played Chinese chess with Chern since. What Yau said is believable because after defeating me he avoided playing Chinese Chess with me again.

The 1970's were especially turbulent for overseas Chinese students.

The Cultural Revolution in China and the Vietnam war raged on.

After years of silence, the community of Chinese students erupted at the Japanese government's seizure of a tiny Chinese island called Diao-Yu-Tai. We students passed out petitions and denunciations. We sought the supports of the Chinese professors and wanted them to take a tough stand. Chern was naturally looked upon to lead. He did not take the same path as the students in protesting but was effective in his own ways. He organized many prominent Chinese scholars to place a full page advertisement in the New York Times protesting the Japanese aggressions. Many students at Berkeley including Paul Yang, Yau and me spent many hours organizing all sort of activities. In the Spring of 1971, Chern had an operation. Afterwards we visited him at the Kaiser Hospital in Oakland. He enthusiastically greeted us and we talked a lot. He advised us to spend less time in student activities and more time concentrating on mathematics. He told us that a good theorem is like a monument, it will exist forever. At that time in Berkeley, this advice was considered to be politically incorrect. I remember that we argued with Chern though he was recuperating from the operation. In general, Chern likes to argue with us. I can tell that he is pleased when he wins the arguments. He and Yau are a good match. It is fun to listen to them argue, about anything ranging from food to Chinese history.

For obvious reasons, I wanted to work in geometry. Yau joked that if I worked with other geometers in Berkeley, I would become only a grandstudent of Chern and hence I would have to pay respect to him, since he was a student of Chern. Though this was a joke, I didn't want it to happen. In 1971, I asked Chern to be my advisor. He agreed, asking me what subject I was interested in. I told him that I was interested in the eigenvalues of the Laplacian. He then told me to give some talks on the work of Minakshisbundaram and Plejel on the heat equation approach to eigenvalues. Chern has many students and gives them a lot of freedom. Though he may not give very specific thesis topics, he is very effective by suggesting a paper to read or another professor to talk to. When I had difficulty in computing the Laplacian on forms for Hermitian manifolds he did several pages of computation and explained to me how to do these computations. Chern was of course very busy and usually our conversations were interrupted by phone calls and visitors. Because Chern got up very early in the morning, I solved this problem by catching him at his office at eight o'clock in the morning. At this time, I could talk to him for an uninterrupted half hour. Chern's administrative skill is well-known. As students our experience was that whenever we asked Chern to write a letter for us or made some other request, he would write it down on a used envelope, but it would always be done on time.

As a student of Chern, I learned much in addition to mathematics. Chern is always very gentle and never forces his opinions on us. On a lot of things I did not heed his advice and I haven't been too successful with those decisions. I hope I'll be smarter in the future. I wish Chern to continue leading a happy and productive life. I think this is the right place to stop, because I need to save something for his ninetieth birthday.

On Being a Chern Student

S.M. Webster
Department of Mathematics
University of Chicago

I was one in the period leading up to June of 1975, when I received my doctor's degree at the University of California, Berkeley. At the time there were very many geometers there – professors, visitors, students. As was commonly said, most were in some sense students of Professor Chern, and my particular experiences were certainly not unique. Whenever I would go to see him there was usually a line of people outside his door, and no one, no matter how important, it seemed, could push in front of even the youngest undergraduate. Once inside, one was constantly interrupted by knocks on the door followed by the brief respects of this or that visitor. With Chern a student had to be quite independent, and I can't remember too many of my mathematical questions being specifically answered.

Yet even after this many years of mathematical activity, I can still say that I have learned more from Professor Chern than any other mathematician. Of course, I was at a very impressionable stage, the transition from student to independent researcher. But of all the interesting, sometimes brilliant, lectures which I attended, those of Chern proved most valuable – the moving frames systematically unlocking the essential geometric secrets (I took up the special case of real hypersurfaces following the work of Chern and Moser); the constant advice to read Cartan, where one could see the most modern and abstract geometric concepts developed naturally with such astounding ease. In addition to promoting the most active and current areas of research, Chern was then and remains a storehouse of all the richness and beauty of the classical geometries. This is so to an extent which is perhaps unique in the Mathematical World.

Shiing Shen Chern, an Optimist

Jean Pierre Bourguignon

Centre de Mathématiques
Ecole Polytechnique, France

I met Professor S.S. Chern a number of times, and, on each of these occasions, I benefited from my contacts with him. He belongs to the small number of people who set an example for their colleagues, young and old. From reading his work and letters, hearing his lectures, and talking with him, I learned many different things, some of them quite subtle.

I also had the privilege of being associated with him in two ventures, one large and one much smaller. The first one was the organization of the meeting initiated by Chern on "*The mathematical heritage of Elie Cartan*" held in 1985 in Lyons. The second one was an interview for the Gazette des Mathématiciens about one year ago.

Meeting Professor Chern

I met Professor Chern for the first time in 1972 at a conference in Oberwolfach. He was then in charge, together with Professor Wilhelm Klingenberg, of organizing the Tagung "Differentialgeometrie im Großen" which takes place every second year at the by now famous German resort. I remember very well the lecture on characteristic classes he gave there, starting (as one should) with his proof of the Gauß-Bonnet theorem by transgressing the curvature form to the unit tangent bundle.

I met him again a year later at the Summer Institute of the American Mathematical Society held on the sumptuous campus of Stanford University. I enjoyed this conference enormously. This meeting was the culmination of the year I spent in the United States visiting Stony Brook at the invitation of Jim Simons. Spending a whole year abroad was a major experience familywise, which my wife and I found so enriching that we decided to repeat it two other times.

The Stanford meeting was decisive for me because it confirmed on a larger scale what I had already learned in Stony Brook, namely, that I was very fortunate to have Marcel Berger as thesis adviser. But, in the mathematical circles in France, and especially those of Paris, the dominant opinion was that Differential Geometry was not a hot subject, and working in this field meant in particular being unable to do the real thing, i.e., Algebraic Geometry. To see abroad so many renowned mathematicians interested in what I was doing under Marcel Berger's supervision made a great impression on the young mathematician I then was. Of particular significance to me was Professor Chern's concern about the small progress I had made with Annie Deschamps and Pierrette Sentenac on the Hopf conjecture on $S^2 \times S^2$ or about the results I had achieved with Shing Tung Yau on Ricci flat metrics on quotients of $K3$ surfaces (S.T. Yau, then in Stony Brook on his first appointment, had not yet proved the Calabi conjecture for them), or even about some calculations I had made at the request of Jim Simons on critical points of the Chern-Simons functional on the space of metrics. (Needless to say, at that time, this object had not reached the ears of any physicist.) All this forced me to realize that the view of the global architecture of Mathematics I had formed earlier in France might be biased.

I met Professor Chern again in Oberwolfach in 1976, and enjoyed there his lecture on webs. I still have in mind the way he expressed his satisfaction at hearing the solution by Sylvain Gallot of the problem of determining manifolds admitting solutions to overdeterminded systems similar to that satisfied by spherical harmonics. "Nice" he said very simply, but a lot was contained in the nuance of his voice.

The next extensive discussion I had with Chern goes back to 1980. It took place in Princeton. He was there on a short visit. During a meal at the cafeteria of the Institute for Advanced Study, I presented to him some of the projects of Arthur Besse. This polycephalic author had just started one of his major entreprises, i.e., writing a book on Einstein manifolds. I mentioned one of our main concerns, namely to show how Geometry has to be approached in many different ways. He replied that the original plans of Bourbaki included grasping mathematics at the same time from two extreme points: from the very general, and from the very particular, a fact I had never heard[3].

The same year, I felt very honoured when Chern invited me to Berkeley while I was staying at Stanford. He extended his hospitality to a private lunch during which he wanted to know more about ideas I was pursuing on vector-valued forms. I was working on an extension to four-dimensional Riemannian geometry of some work on Yang-Mills

[3]Of course, the publications of the University of Nancago belong to this latter spirit. This fake name, mixing Nancy and Chicago, refers to the fact that, shortly after the war, Nancy hosted a number of the Bourbaki members such as Jean Delsarte, Jean Dieudonné, and Laurent Schwartz, and that André Weil was holding a position in the Chicago department, together with Chern.

fields that I had done in collaboration with H. Blaine Lawson. At that time I was stuck at a place that I did not find decisive. It is only later that I recognized how right was the advice given then to me by Chern: "Keep trying".

One year later, I had the good fortune of taking part in the second of a series of symposia organised in China on Differential Equations and Differential Geometry. I understood more of the dimension of Chern while visiting his country. I could then witness how much it meant to him to construct an Institute of a new type at Nankai University. Since then, this Institute has been running with great success.

The mathematical heritage of Elie Cartan.

Over the years, the American Mathematical Society has organized a number of meetings on the mathematical heritage of great mathematicians. Chern felt that one had to be organized to celebrate Elie Cartan's heritage, and that the proper setting for such an endeavor be an American-French cooperation. He was undoubtedly the driving force behind this project, and convinced all colleagues necessary for the success of the conference to join.

The late Edmond Combet and I were enrolled as secretaries of the organizing committee into what revealed itself a major entreprise. The planning of the meeting went very smoothly. Chern had formed a very clear idea of the format that the conference should have, and he had no real difficulty convincing the other members of the organizing committee of the exactness of his views. Almost all proposed speakers accepted at once to participate in the celebration.

Having by now been involved in the organization of other international conferences. I retrospectively measure how exceptional such circumstances are, and I have no doubt that Chern's influence and rightness of views have been decisive in this process. I developed a very effective cooperation with the Mathematics department in Lyons who hosted the meeting. We all knew very well how much it meant to Chern, and we felt an obligation to make the conference an event.

The preparation of the meeting was also for me an opportunity of renewed contacts with Professor Henri Cartan. I could witness his deep friendship with Professor Chern at work. He helped us in all possible ways providing us with pictures, and personal information. I had met him a few years earlier when I was in charge of the reissuing of the Complete Works of Elie Cartan, and had appreciated his commitment to help keep the memory of his father alive. On that occasion, his simplicity and his attention to details impressed me greatly.

The conference was, I think, a great success, developing a life of its own. It attracted a lot of attention, and I could witness the worldwide interest aroused by Elie Cartan's work. It is very well known that Chern has taken all possible opportunities to make this work known, in particular the method of moving frames that Elie Cartan (and Chern) exploited with great success. Browsing through the exposition that Elie Cartan himself made of his work in his *Notice sur les travaux scientifiques*[4] before entering the Academy at age 62 is an experience. But reading Elie Cartan is not always that easy. Chern disagrees on that point. He once said: "After spending a year in Paris (1937–1938), I found Elie Cartan easy to read. I could more or less think in the same way as him". It is then no surprise to see that Chern has been his great continuator, and brought his work to a level of universal recognition that it did not receive in Elie Cartan's life time.

Interviewing Chern.

On Saturday July 14th 1990, on an afternoon of recess of the

[4]which can be found in his Complete Works (the second edition is unfortunately already out of print).

Summer Institute of the American Mathematical Society devoted that year to Differential Geometry, there took place in Los Angeles a very special ceremony to honour Chern. Many of his former students and friends spoke, some of them in a very moving fashion. Aware of this gathering before leaving France, I contacted Henri Cartan. A few days later, he sent me copies of the first two letters that Chern sent to Elie Cartan, and asked me to present them to Chern during the ceremony.

I took the opportunity of Chern's presence at the Summer Institute to do an interview with him for the Gazette des Mathématiciens. This journal, the official publication of the Société Mathématique de France, has inaugurated a few years ago a series of interviews of mathematicians of importance. I approached him the next day during a group visit to the Paul Getty Museum. He was strolling through the magnificent surroundings with his family who had joined to attend the celebration. I was very pleased when he immediately agreed. I was helped in the fine wording of the script by Anthony Philips who offered to shoot a video of the interview. Chern responded very positively to this new and more challenging format for the discussion. The one-hour long video has been edited, and the interview published in the April 1991 issue of the Gazette des Mathématiciens. I refer to this publication for details on the content covering various aspects of his mathematical life (in particular his relationship with Elie Cartan), his feelings about the evolution of the practice of our profession, his vision of the future of contacts between China and the rest of the (mathematical) world. On many of the issues that we discussed, Chern showed an infectious optimism. This led the editor in chief of the Gazette to entitle the interview: "Entretien avec un optimiste, S.S Chern", a title which got Chern's blessing.

Some lessons taught.

From my encounters with Professor Chern, I learned different things. Gaining a certain perspective on my own work early in my career is the one I value the most. Since this view conflicted with the opinion induced upon me by my immediate environment, it forced me to get some distance from my inbred feelings, and for that purpose some is never enough. This achievement revealed itself to be important later on.

Almost equally valuable for me is the even way in which Chern speaks with young and old colleagues. This taught me how critical it is for mature mathematicians to keep contact with the younger generation. Being encouraged by somebody of Chern's stature can mean a lot, and this may make the difference at some point in life. To find the clue sometimes only requires persevering a little more.

Another point has to do with the (disputed) notion of mathematical taste. That such a category exists is sometimes negated although we do make use of it consciously or unconsciously in a number of instances. That Geometry has very many facets is often considered as one of the distinctive features of this field. The work of Chern testifies to the fact that this versatility does not justify publishing papers which are not accomplished or statements which follow from ad hoc assumptions confined to a corner of the field. The number of his articles containing a crucial calculation that later played a decisive role in other subdomains of Geometry is impressive. Their reading helped me shape a view of Geometry in which the interplay between subfields plays a decisive role in its development, and in which methods should never be presented in a way which may later restrict their use to too narrow a domain.

In closing, I would like to express my thanks to both Professor and Mrs. Chern for the lessons of kindness, openness and availability they taught me over the years.

Chern, Some Recollections

Jon Wolfson

Department of Mathematics

Michigan State University

When I entered graduate school at Berkeley in the fall of 1977, I knew of Chern's mathematical fame though I knew little of his work or interests. I first heard him lecture within a couple of weeks of the beginning of the fall term. If I recall correctly he spoke on the sine–Gordon equation and the Bäcklund transformation. It was a beautiful talk – vintage Chern. He began with a review of the method of moving frames, he adapted the moving frame to the problem and proceeded to the main theorem. Throughout he drew schematic pictures to emphasize the geometry of the problem. I and the other new students in geometry were very impressed. In less than an hour Chern had gone from elementary geometry to a beautiful theorem. Moreover, he stressed simplicity throughout. Of course, I heard many more such lectures during my years at Berkeley. Whatever the topic, Chern always emphasized the simplicity of the ideas and the power of the method of moving frames.

I first met Chern personally later that fall term when he had a party at his house for participants of his seminar. He and his wife were gracious and generous to everyone. My wife and I were struck by Chern's humility. In no way is this humility false. Chern is aware of his accomplishments and his place in mathematics. He is justly proud of them. Yet his personality and his insight into mathematics give him the grace to carry his accomplishments with humility. I recall, at another party at his house a couple of years later, Chern discussing mathematics history with a group of graduate students gathered around him. He was emphasizing the theme that all mathematical accomplishments, however magnificent at the time, are but small steps in the whole enterprise.

I formally became a student of Chern in the spring of 1979. As I matured mathematically we became closer until we were working together on various papers. I recall one incident during this period of joint work which, at the time, I knew reflected something of Chern's character. I devised a procedure and named it. I met with Chern and discussed the procedure. He was pleased with the procedure but he clearly did not like the name, though he did not say anything explicitly about it. Rather, he quietly suggested an alternate name. We did not agree. I continued using the name I liked and Chern went along. But he was not happy. Through subtle suggestions on a few appropriate occasions he convinced me of his choice. Quietly he prevailed. This is a simple example of what I have glimpsed often when I have seen Chern dealing with people and situations. Through quiet diplomacy he reasons and convinces. Chern is a master statesman.

In the last few years, fortunately, I have been able to see Chern, and often Mrs. Chern, about once a year. They are always very friendly and kind to me and my family. Invariably when we meet they take me (and my family if they are with me) to a Chinese restaurant in the East Bay. Somehow, each time, regardless of the restaurant, the meal we are served is one of the finest I have ever had. Chern's selections and the restaurant's attempt to please him lead to excellent meals. Now that my family live in Michigan, my wife and I often think wistfully about those meals. Chern and his wife are generous in their invitations. In fact, during the period I was working with Chern I often visited Berkeley and Chern often took me to lunch. So often that the maitre d' in one Chinese restaurant in Berkeley began to recognize me. (It goes without saying that Chern was recognized.) It was a pleasure to later accompany friends to this restaurant and to be greeted with recognition.

My friendship with Chern has been both mathematically and personally a wonderful experience and a highlight of my life.

S.S. Chern, as my Teacher

Shing Tung Yau
Department of Mathematics
Harvard University

I heard about Professor Chern during my fourth year of middle school in Hong Kong through an article published in a popular magazine. The vivid description of the Chinese mathematics community left a deep impression on me. The brief introduction by the editor of the magazine, which made it clear that Chern was a leading mathematician, helped me realize that there is no obstruction for a Chinese to be a world class mathematician. More importantly, by reading Chern's biography, I recognized that thinking mathematics is part of the daily life of a mathematician. Chern's article has therefore opened up the eyes of a young student in Hong Kong.

I was excited by the beauty of mathematics even before I read Chern's article. However, the article was a strong motivation for my commitment to mathematics. My high school teachers and my teachers in College were an inspiration to me. On the other hand, the scope of the interest was limited largely to algebra and analysis. My college teacher, Stephen Salaff, was very excited about the life of Professor Hua in Mainland China. He wrote a biography of Hua. I read a lot of books by Hua, partly because they were the most inexpensive mathematics books and partly because Hua is very popular in Hong Kong. The books by Hua are delightfully well written and contain numerous beautiful computations. Perhaps I was not born to be a number theorist. I did not get the same excitement in reading these books than with those that I read in graduate school by Chern. This is perhaps due to the fact that an exciting and inspirational atmosphere in mathematics only arises when a good crowd of excellent mathematicians are brought together.

In any case, my college teacher Salaff graduated from Berkeley. He felt strongly that I should go to Berkeley for graduate school. Professor Chern was clearly the most influential person for my admission into this famous university. I think it was an unprecedented case for a third year student from Hong Kong to go to graduate school at Berkeley on a fellowship. This means that Professor Chern has made a tremendous effort for me. Prof. Chern is very much willing to help young mathematician once he recognizes their potential. This fact is manifested by many of the articles written in this volume. There were many foreign students, not just Chinese, who felt lonely and found life difficult in the United States during the early stages of their careers. Professor Chern's help was always very timely and the effect often lasted for a long time.

During my first year at Berkeley, I read the book "Morse theory" by John Milnor and was fascinated by its presentation. The brief discussion of geometry with negative curvature impressed me. Immediately, I started to think about all possible rigidity questions related to the beautiful Pressman theorem. I thought about various rigidity problems from different vantage points. In particular, I was fascinated by the relation of harmonic maps to rigidity questions. In fact, I was awarded a grant by Stony Brook two years later for pointing out the relation between harmonic maps and rigidity problems. These ideas have influenced me up until now. During this time, Professor Chern happened to be on leave for a quarter. But he was overjoyed to see my progress. I remember the smile on his face when he learned that I could produce some original work. It is difficult to overestimate the positive influence of such encouragement on a beginner.

I was never satisfied with my work on manifolds with negative curvature. I kept on thinking about complex geometry which I thought had deeper structure. Naturally this line of thought was influenced by Professor Chern's profound work on the subject. In the summer of 1970, I asked him to be my advisor and he kindly accepted. After a month, I was surprised to hear from him that I should graduate with what I had done already although I thought there was still a long way for me to go before finishing graduate school. I went along with what he suggested partly because I hoped to be able to support my family as early as possible. However, Professor Chern has already shown me the beauty of complex geometry, the Chern classes, the secondary characteristic classes, the technique of creating local invariants from geometric structures, and the beauty of projective and affine geometries. Needless to say, each time I visited Berkely, I was truly inspired by discussions with him on mathematics. These influences on my mathematics career are far more important than working on a special problem.

I remember one day when Professor Chern came back from Princeton. He told me that he knew Andre Weil very well and Weil told him it was time for the Riemann hypothesis to be solved. He suggested that I should work on the problem immediately. At this time there was vigorous movement among overseas Chinese students. I spent quite some time with my friends on this movement. Professor Chern was not terribly happy with our excitement. Soon afterwards, he was sick and stayed in the hospital for over a month. We went to see him often in the hospital and I did not spend much time on the problem that he suggested.

Since I was going to graduate, I had to write up my thesis for submission to the graduate school. I wrote a lot in the thesis. However, I did not have a typewriter. I told Professor Chern about it and was surprised when he offered me the use of the beautiful typewriter in his office. In particular, it meant that I had the privilege to use his office when he was not there. I was really flattered by this privilege.

After I graduated from Berkeley, I looked for jobs and did my research. I always consulted him. Once in a while, we had some disagreements, especially ones related to China. There were times when we misunderstood each other. However, there cannot be any question that I have always had a tremendous respect for him. Besides being my teacher, he has been a fatherly figure for me. I owe him for all the advice and help that he gave me. Among all my experiences with Professor Chern, I enjoyed most the time we spent together. I believe everybody shares the same opinion.

I remember I learned from him in 1970 that he thought it is a good contribution to provide interesting open problems for researchers in the field. He provided ten open questions in the Nice congress. As a lesson, I therefore contribute here a set of open problems in geometry. I wish to dedicate them to him for his long life.

Open Problems in Geometry

Shing Tung Yau[5]

As was suggested by Professor Chern, a field is active if it has many interesting open questions. Therefore I have collected some problems of interest to me. I dedicate this paper to Professor Chern on his 79th birthday. I wish to thank my friends S.Y. Cheng, S. Bando, Y. Kawamata, P. Li, S. Nishikawa, and R. Schoen for comments. I also would like to express my deep gratitude to Laura Schlesinger for her patience in typing this manuscript even after I made substantial changes many times.

I. Metric Geometry

1. Find a general method to construct canonical metrics on a compact Riemannian manifold. All the known canonical metrics are obtained by variational principle. In the past, geometers studied the critical points of the functionals defined by the total scalar curvature, the L^2-norm of the Riemannian curvature or these functionals coupled with the Yang-Mills fields. If there is a compact group acting on the man-

[5]Research supported by DOE Grant DEF02-ER-25065 and NSF Grant DMS-89-0822

ifold, one can reduce the variational problem to a lower dimensional problem which may be easier to be solved. The last approach was used by relativists in looking for exact solutions for the Einstein equation. A possible fruitful approach to find canonical metrics is to use singular perturbation method. For example, if two compact manifolds admit Einstein metric with scalar curvature of the same sign, can we connect these two manifolds along certain submanifold so that the sum of the manifold admits an approximate Einstein metric which we can perturb to form an Einstein manifold. Donaldson and Friedman [DFr] has demonstrated this for self-dual metrics where they take connected sum. Recently, Taubes (to appear in Journal of Diff. Geom.) has proved a general existence theorem for self-dual metrics. Since the connected sum of four-dimensional torus does not admit Einstein metric, the problem for Einstein metric is more complicated. Perhaps one has to select the submanifold suitably before we do the perturbation. Would this procedure be much easier for dimension greater than five because there is no known obstruction for these dimensions?

We can generalize this problem to some class of metrics which are more general than Riemannian metric. For example, we can look for Finsler metrics or Riemannian metrics whose connection may not be Riemannian but satisfies other conditions related to the Lagrangian.

2. Find a good topological condition for a manifold to admit conformally flat or projective flat metrics. There are obvious necessary conditions. For example, for conformally flat manifolds, the Pontryagin forms vanish and the universal cover of the manifold can be immersed onto a subdomain of the sphere. What is the structure of their fundamental group? One should study bundles whose structure group is the conformal group. Presumably, there is some kind of stability concept among these bundles. Drawing an analogy with the development in Hermitian Yang-Mills connection. Is it possible to define "stable" conformal bundles so that such bundles with vanishing Pontryagin classes

are flat.

3. Schoen and the author (Conformally flat manifolds, Kleinian groups and scalar curvature, Invent. Math. 92 (1988) no. 1, 41-71.) proved that the universal cover of a compact conformally flat manifold with positive scalar curvature is a domain in S^n. Is there a similar theorem for projective flat structure on a compact manifold? Instead of assuming the manifold to be conformally flat, we may assume that the Weyl tensor is dominated by the scalar curvature. Does similar theorem hold for these manifolds if we replace conformal transformations by quasi-conformal transformations?

4. Berger [B] has classified holonomy groups for Riemannian manifolds. If the metric is not Riemannian but allows different signature, the corresponding theorem of Berger should exist. More importantly, we would like to find a *complete* connection whose holonomy group is the given Lie group. In particular, classify complete Lorentzian manifolds with parallel spinors.

5. Gao and the author [GY] proved that every compact 3-dimensional manifold admits a metric with negative Ricci curvature. There should be no doubt that such a metric exists on every compact manifold with dimension greater than three. Can one say anything about the homotopic type of the space of metrics with negative Ricci curvature? Compact manifolds with negative Ricci curvature do not admit continuous group of isometries. If a finite group acts smoothly on a compact manifold, does the manifold support an equivariant metric with negative Ricci curvature?

6. Let M be a compact manifold with positive scalar curvature. Is it true that M admits a finite cover $\widetilde{M}$ so that, after successive surgery on spheres with codimension ≥ 3, it becomes a fiber space whose fiber

is a compact homogeneous manifold? In view of the work of Stolz [St], this is quite likely to be true. Prove that a compact $K(\pi, 1)$ cannot admit a metric with positive scalar curvature. If the dimension is not greater than four, this was proved by Schoen-Yau [SY].

7. The recent works of Sha-Yang (J.P. Sha and D.G. Yang, Examples of manifolds of positive Ricci curvature, JDG 29 (1989) 95-103) indicate that some form of surgery is possible in the category of manifolds with positive Ricci curvature. Unfortunately, in general, the connected sum of two manifolds with positive Ricci curvature need not admit a metric with positive Ricci curvature. One simply takes two manifolds with non-trivial fundamental groups. The connected sum would have a large fundamental group which is not possible. What happens for simply connected manifolds? Algebraic manifolds with positive first Chern class admit metrics with positive Ricci curvature. Can we do surgery on these manifolds which preserve positivity of Ricci curvature?

8. The well-known Lichnerowicz theorem of vanishing of harmonic spinors for compact spin manifolds with positive scalar curvature implies vanishing for certain characteristic numbers. There is no known obstruction for the differentiable structure of manifolds with positive curvature. Is this an accident? Most likely there are other constraints. Can we find them? There are known obstructions on differentiable structure if the curvature is quarter pinched (see M. Weiss, to appear in Jour. of Diff. Geom.). For four-dimensional manifolds, can we say anything about the Donaldson invariants for manifolds with positive Ricci curvature? It is almost for sure that compact manifolds with positive curvature operator are diffeomorphic to the standard sphere. Can one prove it? Can one prove that compact algebraic manifolds of general type do not admit metrics with positive Ricci curvature?

9. Given an n-dimensional complete manifold with nonnegative Ricci

curvature, let $B(r)$ be the geodesic ball around some point p. Let σ_k be the k-th symmetric function of the Ricci tensor. Then is it true that $r^{-n+2k} \int_{B(r)} \sigma_k$ has an upper bound when r tends to infinity? This should be considered as a generalization of the Cohn-Vossen inequality. If the curvature of the manifold is non-negative, one can consider a similar problem for other invariants constructed from the curvature forms.

10. It is proved by Gary Hamrick and D. Royster, (Flat Riemannian manifolds are boundaries, Invent. Math 66 (1982), 405-413) that the Stiefel-Whitney numbers of a flat manifold are zero. Does a similar statement hold for almost flat manifolds? What about hyperbolic manifolds or other locally symmetric spaces if consider only subset of Stiefel-Whitney numbers?

11. Let M_1 and M_2 be two compact manifolds with positive curvature. Let f be a homotopy equivalence between these two manifolds. Then is it true that $f^* \widehat{A}(M_2) = \widehat{A}(M_1)$ where $\widehat{A}$ denotes the $\widehat{A}$-class of the manifold? A related problem is whether there is only a finite number of diffeomorphism types of compact manifolds with positive curvature which are homotopic to a given manifold with positive sectional curvature.

12. The famous pinching problem says that on a compact simply connected manifold if $K_{\min} > \frac{1}{4} K_{max} > 0$, then the manifold is homeomorphic to the sphere. If we replace K_{max} by the normalized scalar curvature, can we deduce similar pinching theorems? Let M be a compact manifold with nonnegative curvature whose integral cohomology ring is the same as the cohomology ring of an irreducible symmetric space with rank greater than one. Is M symmetric? For rank one symmetric space with nonconstant curvature one can ask the same question if the curvature is quarter pinched. One can ask similar questions for

manifolds with nonpositive curvature. The author (unpublished manuscript 1974) was able to settle the question affirmatively using the Gauss-Bonnet and the signature formula when the curvature is quarter pinched and the dimension is four.

13. Let M be a compact manifold with curvature pinched between two positive constants. Can one estimate the injectivity radius of M from below in terms of these constants and the homotopic types of M? It is well known that this is true for even-dimensional manifolds or if $\dim M = 3$.

14. When does a compact simplicial complex admit a metric with non-positive curvature? This is an interesting question even when the dimension of the complex is equal to two. Can most theorems related to the fundamental group of manifolds with nonpositive curvature be generalized to these spaces? For example, can one define the rank of such spaces? Is there a useful concept of locally symmetric spaces? Can one use topological data to characterize metric complex that is covered by Bruhat-Tits building? If a $K(\pi, 1)$ admits a metric so that the strong isoperametric inequality holds, i.e., for any disk D, the area of D is bounded by C length (∂D) where C is a constant independent of D; is it true that the $K(\pi, 1)$ is homotopic to a simplicial complex with strongly negative curvature?

In a remarkable work, Schoen (see part I of [Sc]) studied the regularity theory of harmonic maps whose target space is a simplicial complex with non-positive curvature. For deeper properties of the regularity theory, he needed stronger assumptions on the complex. His work is essential in understanding the strong rigidity problem for discrete groups in Lie groups (see the work of Gromov-Schoen [to appear]. Is the fundamental group of a compact space with negative curvature "rigid" in some sense? For example, is the mapping class group of the fundamental group finite? There are works due to Gromov, Thurston,

Gerstein-Short and others who study questions related to these problems.

One should remembers that beyond the finiteness of the cohomological dimension of the group, there is no known topological obstruction for a group to be the fundamental group of a complete (noncompact) manifold with non-positive curvature.

15. Let M^3 be an asymptotically flat three dimensional manifold with non-negative scalar curvature. Let Σ be a stable embedded minimal S^2 which minimizes area in the isotopy class of S^2 at infinity. Can one join Σ to the sphere at infinity by a continuous family of S^2's which satisfies a parabolic equation such that the mass of the spheres can be controlled? Geroch and Jang [Ja], for example, proposed the flow $\frac{\partial X}{\partial t} = -\frac{N}{H}$ where N is the normal and H is the mean curvature. This was studied by J. Urbas and C. Gerhard in $\mathbb{R}^3$ [G] and Huisken [Hu] in some nonflat metrics and was used to prove the Penrose inequality, comparing the total mass and the area of Σ for those cases. It is possible that Penrose's conjecture can be generalized to say that for any embedded surface which is isotopic to the sphere at infinity and which encloses the outermost apparent horizon, the Hawking mass is not greater than the ADM mass of the three dimensional manifold. From this point of view, there is perhaps some way to relate the Willmore functional with the Penrose inequality.

16. Let M^3 be a three-dimensional asymptotically flat manifold with zero scalar curvature. If the metric is rotationally symmetric outside a compact set, is it true that the metric is rotationally symmetric everywhere? How much information on M^3 can be drawn from the asymptotic expansion of the metric? Can we find this information effectively?

17. Let M^3 be a three-dimensional asymptotically flat manifold with non-negative scalar curvature. If the total mass is small when we nor-

malize the L^2 norm of the curvature tensor to be one, is M diffeomorphic to R^3 so that the metric is also uniformly equivalent to the flat metric? (It is not clear that our normalization here is natural.) One expects that the manifold admits no stable minimal sphere and perhaps the flow mentioned in problem 15 exists globally which will shrink the sphere to a point.

One can generalize the above two questions to include a more general "Initial data set" formulation in general relativity.

II. Classical Euclidean Geometry.

18. Given a compact surface Σ in R^3 and distribution of distinct points $\{p_1, \cdots, p_n\}$ on Σ, we can attach an integer m in the following way. For a polynomial Q of degree not greater than r in R^3, we can require that

$$\int_\Sigma Q = \sum_{i=1}^n Q(p_i)c_i,$$

where c_i are numbers depending only on p_i.

If r is small, we can require this formula to be valid for all polynomials Q with degree not greater than r. (The c_i's have to be chosen depending only on p_i and r.) The integer m will be the largest integer r chosen in this form. It depends on the distribution of $\{p_1, \cdots, p_n\}$. Keeping n fixed, and letting the p_i's move, we can maximize m and obtain an integer $f(n)$. It would be interesting to study $f(n)$, the corresponding distribution of points and the corresponding weights $c_1, \cdots, c_n$. How unique is such a distribution and can we characterize them geometrically? For example, is it tight in the sense that any small deformation may lead to increase some of the geometrical distances at the same time? Such a distribution may be considered as the most evenly distributed points according to integration. How do we determine the weights c_i?

Instead of points, we can also consider a set of plane sections of Σ and replace $Q(p_i)$ by the integral of Q over the plane sections. In this way we may find "evenly distributed" planar sections of Σ.

The definition defined here depends on the embedding of the surface into R^3. For an abstract manifold, we can replace the space of polynomials with degree less than n by the space of eigenfunctions with eigenvalue less than a fixed constant.

One can find a similar formulation for algebraic cycles in an algebraic manifold M. Let N_i be irreducible algebraic subvarieties with codimension k. Then for each holomorphic section s_i of $\Lambda^k(T_M|N_i)$ and a holomorphic section t of K_M^m, we can take the interior product of $s\bar{s}$ with $(t\bar{t})^{\frac{1}{m}}$ and obtain a (presumably degenerate) volume form over N_i.

For a distinct set of irreducible algebraic subvarieties $\{N_1, \cdots, N_n\}$, we can study the equation

$$\int_M (t\bar{t})^{\frac{1}{m}} = \sum_i \int_{N_i} i_{s_i\bar{s}_i}(t\bar{t})^{\frac{1}{m}}$$

where t is a holomorphic section of K_M^m and s_i are holomorphic sections of $\Lambda^k(T_M|N_i)$ which are chosen to be independent of t.

When m is small, we can make the above equation to be true for all $t \in H^0(K_M^m)$. By changing s_i and maximizing m, we can find an extremal m depending only on $\{N_1, \cdots, N_n\}$. By changing N_i also, we can find an integer $m = f(n)$ which depends only on n. The extremal configuration $\{N_1, \cdots, N_n\}$ should be interesting to study. Even when the N_i are points, the geometric characterization of these points is not clear. In the formulation above, we clearly prefer the case when the canonical line bundle K_M is ample so that K_M^n has many holomorphic sections. Of course we can replace K by some other line bundle L if the manifold and L admit some metric.

19. The Minkowski problem is a problem of determining a closed convex surface in terms of the curvature function defined on the unit sphere

via the Gauss map. It would be interesting to find an efficient algorithm to solve the problem numerically. This is related to the inverse scattering problem. It would be very interesting also to find a suitable generalization of the Minkowski problem to non-convex surfaces. Presumably one needs to understand a multivalued support function and multivalued curvature functions. Can these functions determine the closed surface uniquely? If it is unique, will the result be stable under small perturbations of the data? How does one compute the lines of curvature, the asymptotic lines and the umbilical points in terms of Minkowski data?

20. How does one give a robust description of irregular geometric objects, especially the singular set of geometric objects defined by variational principles. For example, how regular are the singular sets of harmonic maps and minimal varieties? Can one generalize the Douglas-Rado mapping approach to minimal surfaces to the case when the boundary is not a Jordan curve but the one-dimensional skeleton of a complex? One should replace the disk by a simplicial complex with "minimal complexity".

21. Given a surface in R^3, e.g., the face of a human being, what is the most efficient geometric way to describe the "features" of the surface. How does the concept of umbilical points, flat points, critical points of curvature, lines of curvature and asymptotic lines help? For a generic surface, how do we describe the distribution of umbilical points, flat points, or critical points of curvature? How many closed lines of curvature or closed asymptotic lines are there? Can one find more than two closed asymptotic curves on a closed surface with $\int K^+ = 4\pi$?

For each point in S^2, we can issue a ray from the origin. If the ray reflects according to geometric optics, the direction of reflected ray defines a point on S^2. Hence we obtain a map from S^2 into S^2. How much information does this map tell us about the surface? Can we get this information numerically efficiently? There should be interesting dynamics if we iterate this map. If we do not follow geometric optics, we allow rays to be issued from one point and be received at another point. If we know the distances from these two points to the surface, we obtain a map from R^3 to R^3. We would like to recover the surface from this map.

22. Prove that every smooth surface can be locally isometrically immersed into R^3 smoothly and globally isometrically embedded into R^4. C.S. Lin [Li] has made a major contribution for the first problem for surfaces with nonnegative curvature. By slight modification of an argument of E. Poznjak (Isometric immersions of 2-dimensional metrics into Euclidean spaces, Uspechy 28, 47-76), one can prove that any metric on T^2 can be isometrically immersed into R^4. (Lecture notes by the author in Berkeley, 1977).

23. If a closed surface in R^3 admits non-trivial infinitesimal isometric deformation up to any order, does the surface admit non-trivial smooth deformations? (This is true if everything is real analytic, see N. Effimov (Qualitative problems and the theory of deformation of surface, Amer. Math. Soc., Translation 37)). Is there a good theory of isometric deformations similar to the Kurinishi theory? For any n, can we find a closed surface (with no boundary) which admits non-trivial deformation of order n?

24. Can every compact surface be isometrically embedded into a three-dimensional manifolds with nonpositive curvature? (We should also consider three-dimensional manifolds with isolated singularity.) Suppose the second fundamental forms are positive definite. What is the moduli of isometric embedding? We would like to choose these three dimensional manifolds by some variational principle. Hopefully it can be isometrically embedded into an Euclidean space with low dimension.

25. For a generic metric on a compact manifold M^n, what is the moduli space of isometric embeddings of M^n into $R^{\frac{n(n+1)}{2}}$. The local problem was studied in E. Berger, R. Bryant and P. Griffiths, *Characteristic and rigidity of isometric embeddings,* Duke Math. J. 50 (1983) 803-892. A compact Hermitian space Mcan be isometricaly embedded into the Lie algebra of its group of isometries. Is it possible that every Kähler metric over M can be isometrically embedded into the Euclidean space of the same dimension?

26. Given a smooth metric with positive curvature on the disk, what is the condition for a space curve to be the image of the boundary of an isometric embedding of the metric in R^3? If we constrain the boundary curve to be planar, can one give a complete description of the isometric deformations of the disk in R^3? One should study similar questions for annulus and other planar domains. These questions are related to isometric embedding of closed surfaces with $\int K^+ = 4\pi$, where we replace disk by annulus.

27. For most uniqueness theorems for isometric immersions, high regularity of the immersions is assumed. However, experience with isometric deformations show that one should allow low regularity for isometric immersions also. Is $C^{1,1}$ a good assumption for the immersion? For example, under this regularity assumption, will the work of B. Halpern and C. Weaver hold (Inverting a cylinder through isometric immersions and isometric embeddings, Trans. Amer. Math. Soc. 230 (1974), pp. 41-70)? Apparently there is $0 < \alpha_0 < 1$ so that $C^{1,\alpha}$ isometric embeddings are rigid if $\alpha > \alpha_0$ and not so rigid if $\alpha < \alpha_0$. What is α_0?

28. Nash proved that every smooth manifold admits a real algebraic structure. It will be interesting to estimate geometric invariants in terms of the degree of the defining polynomials. For example, given $\epsilon > 0$, and $\delta > 0$, what is the minimal degree of a rational function

to approximate the Green's function $g(x, y)$ over the set $d(x, y) \geq \epsilon$ up to an error given by δ. One can of course ask questions related to the heat kernel, the eigenfunctions or the characteristic numbers.

29. Let M be a compact simply connected three-dimensional Riemannian manifold. Can one find an infinite number of embedded minimal surfaces with fixed genus and unbounded Morse indices? (This is not possible if the Ricci curvature of M is positive.) Pitts and Rubinstein [PR] have asserted that if such an example cannot be found, the famous spaceform problem can be solved affirmatively, i.e., every finite group acting smoothly on S^3 is conjugate to a linear action. A related question is: can one estimate a lower bound of the first eigenvalue of an embedded minimal surface in terms of the ambient geometry only? This is possible when M has positive Ricci curvature and was proved by Choi-Wang, a first eigenvalue estimate for the minimal hypersurfaces, JDG 18 (1983), 559-567.

30. Meeks and the author [MY] proved that a minimal sphere in a three-dimensional manifold which minimizes area in its homotopy class must be embedded. What happens if a minimal sphere of index one is obtained by a minimax argument? One can ask a similar question for disks instead of spheres. Affirmative answers would lead to an affirmative solution of the Poincaré conjecture.

Similarly one should be able to give a new proof of the Smale conjecture (Hatcher's theorem) by studying embedded minimal S^2 which are isotopic to the core S^2 in $S^2 \times R$. It is easy to give a metric on $S^2 \times R$ so that the only nontrivial minimal S^2 is the core S^2. One hopes to understand the minimizing procedure in the space of embedded S^2 to prove that such a space is contractible. It would also be nice to find a canonical flow for a closed embedded curve in R^3 to determine whether a curve is unknotted.

31. Is there a nontrivial continuous family of compact, embedded codimension one minimal hypersurfaces in S^n? In R^3, one can construct continuous families of minimal surfaces by using the concept of conjugate surfaces. Is there any other continuous family of complete minimal hypersurfaces in R^n for $n > 3$? If non-trivial continuous families of embedded minimal surfaces in S^3 do not exist, the famous theorem of Choi-Schoen (Invent. Math. 1983) would imply that for each genus, there are only finite number of compact embedded minimal surfaces of that genus in S^3. The behaviour of the volume and the spectrum of the minimal hypersurfaces is very interesting. Is the set of values of volumes of minimal hypersurfaces closed? It is well known that the volume of the great sphere is the smallest among all compact minimal hypersurfaces. Does the volume of the minimal hypersurface given by the product of the spheres give the lowest value of volume among all non-totally geodesic minimal hypersurfaces? B. Solomon thinks that this is the case. He [S1] has also computed the spectrum of many isoparametric minimal hypersurfaces in S^n. They are basically given by algebraic numbers. Do we expect that the zeta function defined by a compact minimal hypersurface to have special properties similar to those that arise in number theory? Do they satisfy some kind of functional equation? Can one relate the spectrum of a closed minimal surface in S^3 to the spectrum of its conjugate surface? An embedded minimal hypersurface bounds a domain in S^n. Can one relate the spectrum of the domain with the spectrum of the minimal hypersurface? What is the structure of the spectrum of a complete minimal submanifold in R^n? Are there eigenvalues?

32. Find an effective way to construct complete minimal hypersurfaces in R^n or S^n with finite topology and without continuous groups of isometries. The standard Weierstrass Representation and the Reflection Principle has no useful counterpart in higher dimension. R. Hardt and L. Simon constructed many examples by perturbing the

minimal cone over a minimal hypersurface in S^n. Meeks [Me] has proved that complete embedded minimal surfaces with finite topology in a non-trivial complete flat three dimensional manifold must have finite total curvature. It is still not known whether the helicoid is the only non-trivial embedded complete minimal surface with finite genus in R^3 whose total curvature is not finite. Complete minimal surface with finite total curvature can be compactified and their theory is very much related to Riemann surface theory. One should give a description of the moduli space of complete minimal surfaces in R^3 with finite total curvature. Can one find a similar theory in higher dimension?

33. The work of Osserman-Xavier-Fujimoto [O],[X],[Fu] has settled the question of the values of the Gauss map for complete minimal surfaces in R^3. (It is still not known whether the Gauss map of a complete minimal surface with finite total curvature can miss three points.) There is basically no generalization to higher dimension except the beautiful work of Solomon [S2] for all minimizing hypersurfaces with zero first Betti number. Can one find a suitable generalization of Solomon's theorem by weakening the last two assumptions? For $n < 8$, is the image of the Gauss map of any complete non-trivial minimal hypersurface which admits no circle group action and is not product in R^n dense in S^{n-1}? Are there non-trivial complete stable codimension one minimal hypersurfaces in R^n for $4 \leq n \leq 7$?

34. Classify all isoparametric hypersurfaces of spheres. Münzner [M1] [M2] showed that these hypersurfaces must be algebraic of degree 1, 2, 3, 4 or 6. The examples of degree $g \leq 3$ were classified by Cartan [C], all are orbits of orthogonal representations. Orbit examples with $g = 4, 6$ were then classified by Hsiang and Lawson [HL], but inhomogeneous examples with $g = 4$ were subsequently found by Ozeki and Takeuchi [OT] and Ferus-Karcher-Münzner [FKM]. R^8 is the lowest dimension where non-trivial area-minimizing hypercones can occur and

two examples are known: the cones over $S^3 \times S^3 \subset R^4 \times R^4$, and over $2S^4 \times \sqrt{2}S^2 \subset R^5 \times R^3$. Are there others?

III. Partial Differential Equations

35. Many hyperbolic systems in nature provide natural singular sets for geometers to study. For example, the Einstein equations in general relativity form a hyperbolic system with a well-posed Cauchy problem. If one starts with nonsingular data, one may end up with a singular spacetime. One of the most challenging problems is to describe the singularity that arises in this way. The famous "cosmic censorship" conjecture due to Penrose is an attempt to describe such singularities. Another challenging problem is to find an explicit solution for the two-body problem in general relativity. It is important to find an explicit solution to the Einstein equation which admits non-trivial gravitational radiation. In general relativity, many important spacetimes are created which satisfies the field equations approximately. Can one perturb these approximation solutions to exact solutions? The fundamental question is to find a semi-explicit solution of the two body problem in general relativity.

36. The concept of scaling or iteration has played a very important role in the modern theory of dynamical systems. Beautiful pictures were created by studying the dynamical system. Geometric objects with fractional dimension are often created. Do any of these sets occur as the singular set of some naturally defined equation arising from a variational problem?

For example, let Ω be an open domain in S^n which admits a complete conformally flat metric with positive scalar curvature. Schoen and Yau (Invent. Math. 92 (1988) no. 1, 47-71.) proved that the Hausdorff dimension of $S^n \setminus \Omega$ is less than $\frac{n}{2}$. Can one characterize this kind of domain? In case the scalar curvature can be made constant, the closed set $S^n \setminus \Omega$ is the singular set of the natural Yamabe equation. For some results in this direction see the deep work of Schoen.

37. The spectrum of the Laplacian acting on p-forms of a noncompact manifold is in general non-discrete. In most cases, it consists of a continuous spectrum and discrete spectrum part of which may be embedded in the continuous spectrum. The continuous spectrum is more stable. When the curvature of the manifold is bounded, which part of the continuous spectrum is stable when we take coverings of the manifold? Is the global structure of the continuous spectrum stable when we change the metric uniformly? When will bands occur in the continuous spectrum? How stable are they? (These questions should have better answers if the manifold covers a compact manifold. One should also study forms with coefficients in a flat bundle.)

Those eigenvalues that are embedded in the continuous spectrum are much less stable. How can we compute them numerically? When the manifold has bounded curvature and finite volume, can we prove that they have finite multiplicities and the corresponding eigenfunctions have reasonable growth control?

In Schoen-Yau (Invent. Math. 92 (1988) no. 1, 47-71.), the least p ensuring the finiteness of the L^p norm of the Green's function for the conformal Laplacian has important geometric meaning. One should ask the similar question for the Green's form, especially for those manifolds which cover compact manifolds and forms of middle dimension.

38. Let $\{\omega_i\}$ be an orthonormal basis of the L^2-space of one forms of a compact manifold M^n consists of eigenforms of the Laplacian. For all $\epsilon > 0$, the volume of the set $\{x : \|\omega_i\| < \epsilon\}$ defines a function $F_i(\epsilon)$. Let λ_i be the eigenvalues associated to $\{\omega_i\}$. Then $\limsup_{i\to\infty} F_i(\epsilon)$ and $\liminf_{i\to\infty} F_i(\epsilon)$ are interesting functions of ϵ. Are they positive functions?

Can one estimate the behaviour of $\epsilon^{-n} \liminf_{i\to\infty} F_i(\epsilon)$ as $\epsilon \to 0$?

The problem may be easier if we replace ω_i by $\sum_{j=1}^{i} a_j\omega_j$ and average the volume over $\sum_{j=1}^{i} a_j^2 = 1$. We can of course consider similar problem for p forms with $p > 1$.

39. Let M^n be an n-dimensional compact manifold and M' be a compact (Galois) cover of M. Can one estimate the p-th Betti number b_p of M' in terms of the order of the Galois group and the geometry of M? In particular, when does $b_p(M)$ tend to zero as the order of the Galois group tends to infinity? Perhaps this is true if $2p \neq n$ and M has negative curvature.

40. Given a singular cycle C which is not homologous to zero in a complete Riemannian manifold M^n, one can always find a closed form ω whose integral over C is not zero. Can one find this form geometrically so that we can control its behavior at infinity? For example, under what condition can we choose ω to be L^2 harmonic form? When the cycle has more structures such as algebraic cycles, Lagrangian submanifolds, etc., we would like to impose more structures on the forms and ask the reverse question.

41. The spectrum for the Laplacian on the two-dimensional sphere is very rigid. There is a good chance that it may determine the metric completely. Recently S.Y. Cheng proved a beautiful theorem that if the multiplicity of the spectrum of a metric is the same as the multiplicity of the spectrum of the sphere, then all the geodesics of the metric are closed. In other words, the surface must be a Zoll surface. Can the spectrum of a Zoll surface have the same multiplicity of the spectrum of S^2? Can one say something about the multiplicity of the spectrum of surfaces of higher genus. For example, if the multiplicity of the spectrum of a metric on the torus is the same as the one on the square

torus, what can we say about the metric?

42. In [U], K. Uhlenbeck proved that for a generic metric on a compact manifold, every eigenvalue is simple with no multiplicity. Can one give an effective condition on the metric to assure that the metric is generic in this sense? Metrics with a special property are unlikely to be generic. For example, do Einstein metrics have degenerate eigenvalue ? (An approach is to find non-trivial operators which commute with the Laplacian. Characterize these manifolds) Can one describe domains where all eigenvalues are simple? Does there exist a domain with only finite number of multiple eigenvalues? If there is a compact group acting on the manifold, the spectrum clearly cannot be simple. But we can replace the question by understanding whether the representation of the group is irreducible or not. The Weyl estimate tells us the principal term of the asymptotic behaviour of the counting functions for the eigenvalues. Does the error reflect some part of the topology of the manifold? For example, can the torus admit a metric whose spectrum behaves like the spectrum of the sphere after discarding the Weyl term? Does the growth of the multiplicity of λ_i depend on the topology of the manifold? What type of topological information can we extract from the spectrum on functions alone?

43. The study of the geometry of the eigenfunctions for the Laplacian or the Schrödinger operator is very interesting. The size of the level sets and the critical sets of the eigenfunctions needs to be estimated. For the former quantity, there are excellent works due to Donnelly-Fefferman [DF], Hardt-Simon and Dong [Do]. For the latter quantity essentially nothing is known. Another very important related question is the estimate of the maximum norm of the eigenfunctions when we normalize the L^2 norm to be one. For what types of manifolds can the maximum norm be estimated independent of the size of the eigenvalues?

44. In the author's joint work with Singer, Wong and Stephen Yau (An estimate of the gap of the first two eigenvalues in the Schrödinger opertor, Ann. Scoula Norm. Sup. Pisa Cl. Sci. (4) 12 (1985) 319-333), one finds a lower estimate of $(\lambda_2 - \lambda_1)d^2$ for a convex domain where λ_1, λ_2 are the first two eigenvalues and d is the diameter of the domain. What is the best constant for such a lower estimate and is there an extremal domain? Can one find a lower estimate of $\lambda_2 - \lambda_1$ for nonconvex domains or for manifolds with boundary? Let M_ϵ be the manifold defined in the products of n copies of S^3 by $d(x_i, x_{i+1}) < \epsilon$ for all i and $d(x_1, x_n) < \epsilon$. We assume $\epsilon = \frac{1}{n}$. Then it will be interested to estimate $\lambda_2 - \lambda_1$ for the operator $-\Delta + c\left[\sum_{i=1}^{n-1} d(x_i, x_{i+1})^2 + d(x_1, x_n)^2\right]$.

45. Melas (On the nodal line of the second eigenfunctions of the Laplacian on R^2, to appear in Journal of Diff. Geom.) recently proved that the nodal line of any second eigenfunction cannot enclose a compact sub-region of a bounded convex domain. Is there a similar conclusion for higher-dimensional euclidean space? (David Jerison has made an important contribution on this topic recently.) To what extent do these conclusions hold for compact manifolds with boundary? What is the topology of nodal sets of higher eigenvalues? For example, can one find an infinite sequence of eigenfunctions whose nodal domains are disjoint union of cells?

46. It is well known that in 1984, Wente [We] finally solved the old famous problem of H. Hopf on the existence of constant mean curvature torus in R^3. Subsequently Kapouleas (N. Kapouleas, Compact constant mean curvature surfaces in Euclidean three space, JDG 33 (1991) 683-715) constructed examples of constant mean curvature surfaces of higher genus. Is there physical meaning of these surfaces? As Barbosa and doCarmo already observed, these surfaces are not stable in the classical sense. Is it possible that these surfaces are stable in

a suitable sense? What is the moduli space of surfaces with constant mean curvature? In Reilly's proof of the Hopf conjecture for embedded surfaces, he actually proved that the immersion of the surfaces cannot be the boundary of an immersion of a three-dimensional manifold. Hence not every topological type of immersion is permissible. Can one classify these immersions topologically? If we normalize the area of the surfaces to be one, what are the possible values of their mean curvatures?

47. What is the Martin boundary of the universal cover of a rank 1 compact manifold with nonpositive curvature? Note that W. Ballman [Ba], (On the Dirichlet Problem at infinity for manifolds of non-positive curvature, Forum Math. 1 (1989), 201-213) did solve the Dirichlet Problem for these manifolds. For complete, simply connected manifolds with curvature bounded between two negative constants, there is deep work due to M. Anderson and R. Schoen [AS] (Positive harmonic functions on complete manifolds of negative curvature, Ann. of Math. (2) 121 (1985), 429-461). When the manifold is a product, there is work due to A. Freire [F], (J. Diff. Geom. 33, 1991, 215-232). It is still an open question to find a general condition on a complete manifold which guarantees that it supports non-trivial bounded harmonic functions. For example, if the manifold is complete, simply connected with curvature bounded above by a negative constant, can one find non-trivial bounded harmonic functions defined over the manifold? What happens if the manifold is finitely connected but not necessarily simply connected? Clearly we have to assume the volume is infinite.

One can ask similar questions for complete Kähler manifolds with negative curvature. One can study the Silov boundary of these manifolds. However, this is premature as we do not even know how to construct non-trivial bounded holomorphic functions based on the knowledge of curvature and topological data alone.

48. Let M be a complete manifold with non-negative Ricci curvature. Is the space of harmonic functions with polynomial growth finite dimensional when the growth rate is fixed? Does R^n have the maximal dimension of polynomial growth harmonic functions? Since the author's annoucement of this problem in the IMU lecture in 1981, Li and Tam [LT] have done the best work on this problem. When M is Kähler, we want to know whether the algebra of polynomial growth holomorphic functions is finitely generated or not. The generators give rise to a holomorphic map of M into $\mathbb{C}^n$. What do the image and the fiber look like? One can replace functions by holomorphic sections of line bundles with bounded curvature.

49. There has been a lot of success in using Richard Hamilton's flow in the space of metrics. There are several major problems that remain to be studied. When the metric is Kähler, Cao [Ca] proved the global existence in time. The asymptotic behavior of the equation is very interesting and should be related to the complex structure of the manifold. When the manifold is three dimensional, the singularity of the flow is perhaps related to the pinching off of the topology of the manifold.

50. R. Bartnik (Comm. Math. Physics 94, 155-175 (1984)) demonstrated the existence of maximal slice under certain coordinate conditions at infinity and in the interior. Assuming only geodesically complete and an asymptotically euclidean Cauchy surface, will such maximal foliation exist? This is possibly related to the splitting theorem recently proved by Eschenberg, Galloway, Baum, Ehrlich, Markvorsen, and Newman. (J.H. Eschenberg, The splitting theorem for spacetime with strong energy condition, JDG 27 (1988) 477-491, R. Newman, A proof of the splitting conjecture of S.T. Yau, JDG 31 (1990) 163-184.) In the statement of the splitting conjecture, can we replace the assumption of the existence of a line by the existence of a complete

stable geodesic.

51. In general relativity, there is a constraint system of equations involving an asymptotically flat metric tensor and another symmetric tensor. There are four constraint equations and it is therefore underdetermined. Can one "solve" this system in an efficient way?

Recently R. Bartnik (to appear in *Journal Diff. Geom.*) has found a way to parametrize a large class of asymptotic flat metrics with nonnegative scalar curvature on R^3. He assumes that there is a foliation by round spheres of increasing area. This last assumption precludes the existence of minimal spheres. It would be nice to give a complete classification with no assumption. More generally, we have another symmetric tensor p_{ij} on R^3 which decays quadratically at infinity so that $\frac{1}{2}\{R-\sum p_{ij}^2+(\sum p_{ii})^2\} \geq \left[\sum_i(\sum_j p_{ij,j}-\sum_j p_{jj,i})^2\right]^2$. Can one parametrize the set of asymptotic flat metrics coupled with this tensor p_{ij} also? It will also be important to parametrize these initial data sets so that trapped surfaces exist? Is there a variational method to prove existence of trapped surfaces? Schoen-Yau (Comm. Math. Physics 90, 575-579 (1983)) found sufficient conditions in terms of matter density. It is important to find conditions involving gravitational content also.

IV. Kähler Geometry

52. Prove that every compact almost complex manifold with dimension ≥ 3 admits an integrable complex structure. This question is still unsolved for the six-dimensional sphere. One approach is to form a parabolic flow in the space of almost complex structures to deform an almost complex structure to an integrable one. Can one give an estimate for the number of components of the moduli space of complex structures over a manifold in terms of its topological invariants?

53. Given a complex manifold, can one find a procedure to recognize whether it admits a Kähler structure or not. All the known invariants come from the Hodge structures. Are there torsion invariants? If a manifold admits two complex structures with the same Chern classes and if one of the complex structures admits a Kähler Einstein metric, is the other complex structure also Kähler? (Note that Hironaka has constructed counterexamples if the manifold is not Kähler Einstein.)

54. The Chern classes of a holomorphic bundle over an algebraic manifold can be represented by algebraic cycles. Is this the only obstruction for a complex vector bundle V to admit holomorphic structure if we allow to take direct sum of V with some trivial bundle? What is the obstruction for cohomology classes of a compact manifold M to be representable as Chern classes of holomorphic vector bundles over some complex manifolds diffeomorphic to M? Obviously they are integral and they have to be of type (k, k) for some complex structure. If we restrict the holomorphic vector bundle to be natural bundles generated by the tangent bundle, the obstruction being of type (k, k) is certainly not enough. What are the further criteria? For example, when will an integral class of (k, k) type be the Chern class of some complex structure of the manifold? For complex surfaces, this is related to the question of the distributions of C_1^2 and C_2.

55. If a compact complex manifold M can be deformed to a Kähler-Einstein manifold with non-positive scalar curvature, then is M Kähler? A special case is the $K-3$ surface. Todorov had an idea of using the surjectivity of the Period map. Based on the author's suggestion of using certain special Hermitian metric, Siu was able to complete the proof for the $K-3$ surface. Perhaps generalization of these ideas can be used in higher dimension.

56. Prove that a compact Kähler manifold can be deformed to an

algebraic manifold. Is it possible that every compact Kähler manifold can be embedded into some (singular) universal object which can be deformed to $\mathbb{CP}^n$?

57. The algebraic dimension of a compact complex manifold is the dimension of the field of meromorphic functions. When it is equal to the dimension n of the manifold, the manifold is called Moishezon. It is proved by Moishezon that after blowing up along subvarieties, a Moishezon manifold becomes algebraic. Can one control what kind of subvarieties for blowing up? It often happens that there are complex cycles in a Moishezon manifold which is homologous to zero. Is it a general phenomenon? If the algebraic dimension is $n-1$, then Ueno (Springer Lecture notes 439) proved that it is birational to an elliptic filter space over a Moishezon manifold. What is the corresponding statement when the algebraic dimension is smaller?

58. Fix a complex manifold M, what is the moduli space of quasiprojective structures over M with the given complex structure. Characterize those M where the dimension of the moduli space is non-zero. Can one determine those birational invariants of the quasiprojective structure which depend only on the complex structure? Is it possible to determine the number of components of the moduli space in terms of these birational invariants if we identify quasiprojective structures by birational maps which is regular on M? What is a good condition for the number to be equal to one? (This is the case when the first Chern class of M is negative.) For M to be of low Kodaira dimension, the problem is more difficult. In particular, it is not known whether every projective compactification of $\mathbb{C}^n$ is rational or not.

There are corresponding questions related to holomorphic vector bundles. Can one find intrinsic conditions on holomorphic vector bundles over a quasiprojective manifold so that they can be extended to certain compactification of the manifold? If two algebraic vector bundles are biholomorphic to each other on a quasiprojective manifold, when are they algebraically equivalent? The famous Serre conjecture is the special case when the manifold is $\mathbb{C}^n$.

59. Let M be a compact complex manifold. Then we can form a subgroup of the K-group of topological vector bundles by those elements representable by holomorphic vector bundles. The relative position of these two groups should tell us information about deformation of complex structures over M. When M is algebraic, algebraic cocycles are the images of the subgroup under the Chern character. The Hodge conjecture says that this corresponds to closed forms of (p,p) type which are integral. What kind of bundles represent closed forms of (p,q) type under the map of the Chern character?

Let M be an algebraic manifold. Let $K_{alg}(M)$ be the set of all algebraic vector bundles over M modulo the following relation. We say that two algebraic vector bundles V_1 and $\widetilde{V}_1$ are equivalent if there are algebraic line bundles L_i and algebraic vector bundles V_j so that $\oplus(L_i \otimes V_i)$ is algebraically equivalent to $\widetilde{V}_1 \bigoplus_{i>1} V_i$. Under this equivalence, $K_{alg}(M)$ becomes a ring. How well does it determine the algebraic structure of M?

60. Given an algebraic manifold M, what is the condition on a divisor D with normal crossing so that the complement is a $K(\pi, 1)$ or admits a complete Kähler metric with negative curvature? Can one compute the moduli space of the quasiprojective structures over the complement in terms of the topology of (M, D) and the moduli space of M?

61. Recently, in independent works, Looijenga (E. Looijenga, L^2-cohomology of locally symmetric varieties, Comp. Math. 67 (1988), 3-20.) and Saper-Stern (L. Saper and M. Stern, L^2-cohomology of arithmetic varieties, Ann. Math. 32 (1990) 1-69) settled the famous Zucker conjecture. For a large class of locally symmetric spaces with fi-

nite volume, the L^2-cohomology is shown to be isomorphic to the intersection cohomology of the space with the Baily-Borel compactification. When does a complete manifold with finite volume and non-positive curvature have finite dimensional L^2 cohomology? In case it does. can one find a suitable compactification whose intersection cohomology is the L^2 cohomology?

When a manifold is Kähler, we may have a better answer. Perhaps a complete Kähler-Einstein manifold with finite volume always has finite-dimensional L^2 cohomology which is isomorphic to the intersection cohomology of a suitable compactification. Perhaps the compactification can be described by the ring of L^p-holomorphic sections of pluricanonical line bundles for all $p > 0$. Is this ring finitely generated? Tsuji is making important contributions on this problem.

62. Let M be a complete Kähler manifold which admits a uniform Lipschitz diffeomorphism onto the Euclidean space with the same dimension. Is M biholomorphic to the complex Euclidean space. This can be considered as a "parabolic" uniformization of higher dimensional Kähler manifolds. One can replace the Euclidean space by other models. For example, if the Euclidean space is replaced by the total space of a holomorphic bundle over a compact Kähler manifold, one can ask whether the Kähler manifold is a holomorphic vector bundle with the same fiber dimension.

63. Let M be a complete Kähler manifold whose volume growth is not greater than the euclidean space with the same dimension. Are there non-trivial bounded holomorphic functions on M? How about polynomial growth holomorphic functions? Do they form a finitely generated ring?

64. Let M be a compact simply connected Kähler manifold with zero Ricci curvature. Is it possible to classify all stable minimal S^2 in M?

Are they holomorphic with respect to some complex structure over M? It would also be interesting to settle the question of area minimizing surfaces in R^n. The work of Micallef [MM] has solved a large part of this question.

65. Prove that a compact Kähler manifold with positive first Chern class admits a Kähler Einstein metric iff the manifold is stable in the sense of geometric invariant theory, the tangent bundle is stable as a bundle and the automorphism group is reductive. The most significant result known so far is due to G. Tian [T1] and [T2]. Although a few papers were published after the works of Tian, substantial ideas will be needed to go substantially beyond Tian's results already achieved.

66. Let M be an n-dimensional compact Kähler manifold with non-negative first Chern class. Let ω be an integral Kähler class. Can one estimate $c_2 \cup \omega^{n-2}$ from above in terms of ω^n. Is $c_2 \cup \omega^{n-2}$ always positive unless M is covered by the torus? Are there only a finite number of deformation classes of manifolds with non-negative first Chern class for each dimension? If M is not covered by the torus, can one always find a rational curve in M? Of course, the theory of Mori [Mo] and Wilson [Wil] tells us a lot about this.

67. Let M be a compact Kähler manifold with positive holomorphic sectional curvature. Is M unirational? Does M have negative Kodaira dimension? It is not even clear that M is algebraic.

It is well known that this class of manifold must be simply connected because the second variational formula for closed geodesics shows that a closed curve can always be shortened. When M is a surface, Hitchin [H] proved that this class of manifold is exactly the class of rational surfaces. If an algebraic manifold is obtained by blowing up a compact manifold with positive holomorphic sectional curvature along a subvariety, does it still carry a metric with positive holomorphic sectional curvature?

It would also be interesting to find a good sufficient condition for an algebraic manifold to support a Kähler metric with negative holomorphic sectional curvature.

68. Find a general geometric criterion for an algebraic manifold to be unirational. For example, which Fano manifolds are unirational? Can we find a differential geometric criterion to distinguish the concept of unirationality and rationality?

69. Let $\widetilde{M}$ be the universal cover of a compact algebraic manifold M of general type. Suppose $\widetilde{M}$ has finite topology and contains no rational curves. Is $\widetilde{M}$ bimeromorphic to a subdomain of some algebraic manifold N so that the covering transformation group of M is represented as a subgroup of the group of birational transformations of N? If M contains rational curves, then we should look at the universal cover of the complement of the subvariety which contains all the rational curves. If $\widetilde{M}$ is a bounded domain in $\mathbb{C}^n$ and if $b_1(M) = 0$, would the moduli space of M be compact? Compact Kähler manifolds with $\Pi_i = 0$ for $i > 1$ tend to be rigid. Can one describe this rigidity? Can one classify Π_1 of these manifolds? If Γ is a discrete subgroup of the group of isometries of a Kähler manifold $\widetilde{M}$ such that $\widetilde{M}/\Gamma$ is compact, is Γ isomorphic to the fundamental group of some compact Kähler manifold? This is the case if Γ preserves a Hodge metric and if it admits a torsion free subgroup with finite index.

70. Let M be an algebraic manifold whose tangent bundle is semi-positive. If the Kodaira dimension of M is negative, is M homogeneous? If the Kodaira dimension of M is zero, is M covered by the torus? When dim $M = 3$, Zheng [Z] was able to give an affirmative answer. Campana told the author that Peternall and he can prove the same statement. (F. Campana and Th. Peternall, Projective mani-

folds whose tangent bundle are numerically effective, Math. Ann. 289 (1991) 169-187.)

71. Let M be a complete Kähler manifold with positive Ricci curvature. Prove that M is a Zariski open subset of some compact Kähler manifold.

72. Kähler manifolds can be considered as symplectic manifolds. The interplay between the symplectic structure and the complex structure should be very interesting. For example, Schoen and Wolfson have shown that the area of a closed Lagrangian surface in a compact two-dimensional Kähler manifold can be minimized in its homotopy class to obtain a minimal Lagrangian surface. When the Kähler surface is a K-3 surface with a Ricci-flat metric, such surfaces must be holomorphic for some complex structure on the K-3 surface. Is this true for a general Kähler surface with Kähler-Einstein metric? In particular, does every simply connected Kähler surface admit a complex structure which supports a rational curve?

When two Kähler manifolds are symplectically diffeomorphic, what can we say about their complex structures? It takes some effort to construct a four-dimensional symplectic manifold which admits no Kähler structure. Besides the inequality $3C_2 \geq C_1^2$, what are the other topological obstructions for a simply connected symplectic manifold to be Kähler?

73. Can one characterize algebraic geometrically those algebraic manifolds whose first Chern classes can be represented by a non-positive semi-definite form. This is related to the abundance conjecture which says that for a "minimal" algebraic manifold, some high power of canonical line bundle is generated by global sections. (See Kawamata, Invent. Math. 79 (1985), 567-588). Kawamata also proved that if the canonical line bundle is numerically effective and big, then a high multiple of it

has no base point. This is certainly related to the above problem. Can one prove that for some integer N depending only on the dimension, the N-th multiple of the canonical line bundle gives rise to a birational map into the projective space if the canonical line bundle is numerically effective and big.

74. Let M be a Moishezon manifold with Kodaira dimension equal to zero. Is it true that M is birational to some algebraic manifold (with mild singularity) whose canonical line bundle is trivial? Recently, Kawamata proved this statement for threefolds. Given a threefold with trivial canonical line bundle, can we classify all the other threefolds with trivial canonical line bundle which are birational and diffeomorphic to each other?

75. Can one characterize algebraic geometrically those algebraic manifolds whose first Chern classes can be represented by a positive semi-definite form? If the canonical line bundle is not torsion, does the anti-canonical line bundle have a non-trivial section?

76. Let M be a compact three-dimensional Kähler manifold with zero first Chern class. String theorists have defined a concept of the mirror manifold of M. It is supposed to be another compact three-dimensional Kähler manifold with zero first Chern class. However the Euler numbers should be the negatives of each other. If the second Betti number of one manifold is equal to one, there is a way to count the number of rational curves of fixed degree in this manifold in terms of the moduli of complex structure of the other manifold. Can one formalize any of these concepts in a mathematical manner and justify some of the formulae? It is apparent that the loop space of M or the space of maps from S^2 into M is related to that of the mirror manifold. The moduli space of M is apparently the quotient of the tube domain defined by the Kähler cone of the mirror manifold by a discrete group. In any

case, it is not known whether the moduli space is quasiprojective or not.

77. Let M be a compact three-dimensional algebraic manifold with zero first Chern class and zero first Betti number. Can one find a stable bundle which is a deformation of the tangent bundle plus a trivial line bundle so that the restriction of the stable vector bundle to any rational curve is non-trivial? What is the smallest possible value of $|c_3(V)|$ for a rank 3 stable vector bundle V over the manifold M so that $c_1(V) = c_1(M) = 0$ and $c_2(V) = c_2(M)$. Can $|c_3(V)| < |c_3(M)|$? In general, one would like to construct stable bundles V with $c_1(V) = 0$, $c_2(V) = c_2(M)$ and $|c_3(V)| = 6$ and also the restriction of V to any rational curve is non-trivial. Let $\mathcal{M}$ be the moduli space of stable bundles with $c_1 = 0$, $c_2 = c_2(M)$ and whose restrictions to rational curves are non-trivial. These bundles are special in string theory [Wi]. Can one say much about this space?

78. It is useful in string theory to find those algebraic threefolds with $c_1 = 0$, euler number equal to ± 6 and whose fundamental group is non-trivial. Tian-Yau [TY] constructed several examples by taking a free Z_3 action on a complete intersection of hypersurfaces in the product of projective spaces. All of them are either diffeomorphic to each other or constructed by small resolutions or by taking the orbit space of these manifolds. Subsequently, this construction was used to construct thousands more examples [CGT]. However, no essentially new examples with euler number equal to ± 6 and non-trivial fundamental group were produced. It would be interesting to see if there are some more examples.

79. In order to understand analytically the mirror symmetry, it is interesting to look for Hermitian metrics of the following form. Locally there should be a $(1,0)$ form ω_α so that $g_{\alpha,\bar{\beta}} = \omega_{\alpha,\bar{\beta}} + \omega_{\bar{\beta},\alpha}$ and we

demand $\sum_j \left(\Gamma^j_{ij} - \Gamma^{\bar{j}}_{i\bar{j}}\right) = 0$. What is the condition on a compact complex manifold to admit such Hermitian metrics and how to parametrize them?

80. Donaldson's invariants are constructed by studying the moduli space of self-dual connections over a compact four-dimensional manifold. For algebraic surfaces, can one express the Donaldson invariants in terms of some readily computable algebraic geometric invariants? For example, can one compute them in terms of the pluricanonical ring? For complex manifolds with dimension greater than two, one can replace self-dual connections by Hermitian Yang-Mills connections. One can then define similar invariants. What can we say about them?

81. Let M_1 and M_2 be two complete Kähler manifolds with bounded curvature and finite volume. Suppose M_1 and M_2 are (properly) homotopic to each other. Can one find a homotopy equivalence between M_1 and M_2 with bounded energy. This is important in studying the rigidity problems [J-Y]. If the answer to the question is negative, it would be interesting to classify the topologies of those manifolds which do not support these homotopy equivalences. The same question is not known for all locally symmetric space with finite volume.

82. Let M be a complete Kähler manifold which is a Zariski open subset of some compact manifold. Suppose we want to study deformations of M which can also be compactified. Can one generalize the Kuranishi theory to represent these deformations in terms of L^2 forms in $H^1(T)$. Obviously some condition on the Kähler metric is required. What are these conditions? Y. Kawamata [K] has developed a theory of log deformation which dealt with the algebraic aspect of the theory. However, it seems that more works need to be done on the analytic side. An interesting example can be described as follows. Let M be a complete Kähler surface with zero Ricci curvature. Let ω be the Kähler

form and Ω be a holomorphic two form so that $\Omega \wedge \bar{\Omega} = \omega \wedge \omega$. A linear combination of $\Omega, \bar{\Omega}$ and ω will define a new complex structure over M. How does one describe this complex structure?

83. Find a criterion for a domain in $\mathbb{C}P^n$ to support a complete Kähler metric with one of the following conditions: (1) positive total scalar curvature, (2) positive scalar curvature, (3) positive Ricci curvature, (4) positive holomorphic curvature, (5) positive bisectional curvature, (6) strongly negative Ricci curvature, (7) strongly negative holomorphic sectional curvature, (8) strongly negative bisectional curvature. For the first five cases, one wants to study the complement of the domain. Does it have small Hausdorff dimension? Is it a union of complex subvarieties?

84. There is a beautiful theorem of Belyĭ [Be] that an algebraic curve is defined over the algebraic closure of Q iff it can be branched over P^1 with at most three branch points in $\mathbb{P}^1$ [Belyĭ, G.V., On Galois extensions of a maximal cyclotomic field, *Mathematics of the USSR*, Vol. 14, 1-3, 1979, 247-257]. Can one find a generalization of this theorem to higher dimensional algebraic manifolds? For example, the quotient of a Hermitian symmetric domain can be characterized algebraic geometrically (see [Y1]). Can one distinguish those which are defined over $\bar{Q}$ in terms of algebraic geometric data? It is important to give a purely algebraic geometric proof of the author's theorem that the quotient of the ball is characterized by the Chern numbers. (There is also algebraic geometric characterization of the quotient of other Hermitian symmetric domains by the author also.) Hopefully such a proof may single out those which are quotients of arithmetic groups.

85. Give an algebraic geometric criterion for a holomorphic vector bundle V to admit a flat connection, i.e., the bundle which arises from a representation of $\pi_1(M)$ into $GL(n, \mathbb{C})$. Naturally $c_i = 0$ for all i

is a necessary condition. If it arises from a representation of $\pi_1(M)$ into $U(n, \mathbb{C})$, then by the theorem of Donaldson [D] and Uhlenbeck-Yau [UY], $c_1 = c_2 = 0$ and the stability of the bundle is necessary and sufficient. By putting a Higgs' structure on V, Simpson [Si] has obtained a similar criterion. One should also mention works of Corlette [Co]. After we know the bundle is obtained by representation of the fundamental group, can we detect the character of the representation in terms of algebraic geometric data. Projective flat structures turn out to be very natural also. They occur in relation to questions which arise in string theory. How can they be characterized algebraic geometrically? What happens if the manifold is quasiprojective?

86. Given two locally irreducible Hermitian symmetric space of non-compact type M_1 and M_2. If the moduli spaces of flat vector bundles over M_1 and M_2 are isomorphic to each other for each fiber dimension, is M_1 isomorphic to M_2?

87. If the holonomy group of a compact Hermitian manifold can be reduced to a proper subgroup of $U(n)$, can we say something non-trivial about the manifold. The problem is that the connection need not be Riemannian. For example, if the holonomy group is discrete, the manifold is covered by a complex Lie group holomorphically. It is also very interesting to study the same question for Hermitian connections over holomorphic bundles constructed from the tangent bundle, e.g., the direct sum of the tangent bundle with the trivial bundle. One should also allow the holonomy group to be noncompact.

88. Let M_t be a holomorphic family of compact algebraic manifolds over the disk so that for $t \neq 0$, M_t admits a Kähler Einstein metric. How does the Kähler Einstein metric change and how does the Green's function and the spectrum move when $t \to 0$? Ji [J] and Wolpert [Wo] studied this problem for Riemann surfaces. Jorgenson [Jo] and Went-

worth [W] also studied a related problem when the metric is induced from the Jacobian.

89. For quite a few manifolds (with singularities) that appear as exact solutions in general relativity, one can "Wick rotate" [P] the metric to obtain nonsingular complete Riemannian manifolds with zero Ricci curvature. Is this an accident or is there a useful principle behind this construction? Can one go from complete Kähler manifolds back to physically interesting space times?

90. Let M^n be a compact algebraic manifold of general type. Can it be diffeomorphic to an algebraic manifold which is not of general type? Is the moduli space of algebraic structure over a compact algebraic surface of general type connected? What type of fundamental group can arise as the fundamental group of an algebraic surface of general type? In a recent work, Toledo [To] gave an example of a compact algebraic surface whose fundamental group is not algebraic. Can one characterize those quasiprojective manifold which is the moduli space of an algebraic surface of general type?

91. Let M be an algebraic manifold and let G be a finite group acting smoothly on M. Is it true that G must preserve some complex structure on M. "Exotic" actions should be rare. How to describe them?

92. Let X be a compact algebraic variety. Which part of the topology of X is birational invariant? Let D be any effective divisor which is the direct sum of distinct irreducible subvarieties. There is a partial order on the set of D's by inclusion. By taking limit of various analytic objects on $X \setminus D$, one can define birational invariants of X. For example, one can take limits of $\pi_1(X \setminus D)$ or $H^i(X \setminus D)$. How can these be computed? For the former group, it would be interesting to study its representation theory. If $X \setminus D$ is covered by a noncompact

manifold $\widetilde{X}_D$ with covering transformation group Γ_D, one can study the L^2-cohomology of $\widetilde{X}_D$ as a Γ_D-module. Can one take the limit of these modules to provide birational invariants of X?

93. Classify those compact complex manifolds which contain Zariski open sets isomorphic to Ω/Γ where Ω is an open domain in a compact Hermitian symmetric space and Γ is a subgroup of biregular transformations of this space. Presumably all surfaces of class ∇II_0 appear in this form.

We expect to see a large class of compact complex non-Kähler manifolds with dimension ≥ 3. Hodge theory fails for these manifolds in general. What type of analytic invariants can we introduce to study their complex structures? A Hermitian metric ω on M^n which satisfies the condition $\partial\overline{\partial}(\omega^{n-2}) = 0$ seems to be particularly interesting for many analytic arguments to be useful. It was pointed out to me by F. Zheng that holomorphic one forms must be closed over these manifolds. Can one find a useful criterion for the existence of these metrics? For non-Kähler manifolds, can one find a good concept of canonical Hermitian metric or canonical connection with holonomy group in $GL(n, \mathbb{C})$?

94. Tian [T3] and Todorov have proved that the deformation of compact Kähler manifold with zero first Chern class is unobstructed. Is there a theorem of this type for holomorphic vector bundles? Can one find a similar theorem for surfaces of general type with additional conditions? For example, will the moduli space be unobstructed if the tangent bundle is negative? Many years ago, when the author proposed the rigidity of complex structures on compact Kähler manifolds with negative curvature, the author also proposed the study of rigidity of the holomorphic tangent bundle of these manifolds. Is it possible to find a sufficient condition for a bundle to be rigid in terms of its curvature?

95. Let M be a compact algebraic manifold with ample canonical line bundle. Then the author [Y] proved that M can be written as the quotient of non-trivial product manifolds iff TM splits holomorphically. Most likely such a theorem is true if M is of general type and minimal. Perhaps there is a formulation of such a theorem in terms of the function algebra. In other words, can one find some "linear" characterization of a function algebra to be a non-trivial tensor product or "twisted" tensor product of subfields? Furthermore can one find a criterion for an algebraic manifold to be a nonsingular fiber space over another manifold in terms of its tangent bundle? Clearly the tangent bundle must be the extension of a holomorphic subbundle by another bundle.

If an algebraic manifold with negative first Chern class is diffeomorphic to a product manifold, can the algebraic manifold be written as a non-trivial holomorphic fiber space?

96. Prove that any algebraic manifold that is homotopic to a compact Hermitian symmetric manifold is biholomorphic to a fiber product of these manifolds. What happens if we only assume that the manifold is Moishezon. For the last question, the answer is affirmative if the manifold is three-dimensional. (See Kollár, Survey in differential geometry, 1991, p. 175).

97. A lot of concepts in complex geometry (hyperbolic in the sense of Kobayashi and Brody) have their counterpart in Riemannian geometry. Let M be a manifold (or a simplicial complex). Then M is defined to be topologically hyperbolic if there exists a metric on M, so that there is no distance decreasing conformal harmonic map from R^2 into M. By the theorem of Sacks-Uhlenbeck, it is easy to see that M is a $K(\pi, 1)$. By the recent work of V. Bangert and V. Schroder (Existence of flat tori in analytic manifolds of nonpositive curvature, 24 Ann. Scient. Ec. Norm. Sup. (1991) 605-634), we also expect that if a compact manifold

with non-positive curvature admits no non-trivial abelian subgroup in its fundamental group, the manifold is topologically hyperbolic. Given a compact $K(\pi, 1)$ which admits no non-trivial abelian subgroup in its fundamental group, is the manifold topologically hyperbolic? What would be a good way to generalize the concept if the $K(\pi, 1)$ is not finite dimensional. (Note that if the boundary of the manifold is not empty, we should require the boundary to be convex.)

An analogy of "measure hyperbolic" is the statement that the Gromov volume of the manifold is non-zero. An analogue of an open problem in complex manifold is the following question: Is a compact n-dimensional manifold with non-zero Gromov volume quasihyperbolic. Quasihyperbolicity is a modification of the above definition by allowing the image of the maps from R^2 to be a subset of a fixed lower dimensional subvariety.

Can a compact manifold with nonnegative scalar curvature be topological quasihyperbolic or topological measure hyperbolic? How do these concepts change if we modify the manifold by surgery on embedded spheres with codimension ≥ 3?

98. Let M^n be a complex manifold. Then one can define a holomorphic invariant of M by taking the smallest number of holomorphic maps from $\mathbb{C}^n$ into M^n so that M^n can be covered by the union of these maps. The number is defined to be infinity if we cannot cover M^n by the maps. This number decreases under surjective holomorphic map. Can one compute this number for rational manifolds? What happens if in the above definition, we require the maps to be immersed or embedded? If we look at maps from $\mathbb{C}^k$ into M^n for $k < n$, we can look for the number of irreducible families of such holomorphic maps.

99. For a Kähler manifold of one dimension, the Green's function is a powerful tool and because of the conformal invariance of the Laplacian, it is well understood. In higher dimensional Kähler manifold, we look

for functions φ so that $\omega + \partial\overline{\partial}\varphi \geq 0$ and $(\omega + \partial\overline{\partial}\varphi)^n$ is given by some multiple of the delta function. Can one make use of this non-linear Green's function to produce bounded holomorphic functions? One can also study a linear Green's form by looking at the Laplacian acting on $(n, 0)$ forms.

100. In Riemannian geometry, there is a close relationship between the spectrum of the Laplacian and the lengths of closed geodesics. Are there analogues of those formulae for the Laplacian acting on forms? For Kähler geometry, one would like to study similar formulae for Laplacian acting on (p, q) forms. Do the areas of the minimal spheres relate to the spectrum of some kind of operators acting on the loop space of the manifold?

References

[AS] M. Anderson and R. Schoen, *Positive harmonic functions on complete manifolds of negative curvature*, Ann. of Math. (2) 121 (1985), 429-461.

[B] M. Berger, *Remarques sur le groupe d'holonomie des variétés Riemannians*, C.R. Acad. Sci. Paris 262 (1916), 1316-1318.

[Ba] W. Ballman, *On the Dirichlet Problem at infinity for manifolds of non-positive curvature*, Forum Math. 1 (1989), 201-213.

[Be] G.V. Belyĭ, *On Galois extensions of a maximal cyclotomic field. Math. of USSR*, vol. 14 (1-3) (1979), 247-257.

[C] E. Cartan, *Sur les familles remarquables d'hypersurfaces isoparamétrique dan les espaces sphériques*, Math. Zeit. 45 (1939), 335-367.

[CGT] P. Candelas, P. Green and T. Hübsch, *Connected Calabi-Yau Compactification in "Strings '88"*, p. 155, (ed. by Grates, Preitschopf, and Siegel), World Scientific, Singapore, 1989.

[Ca] H.-D. Cao, *Deformation of Kähler metrics to Kähler-Einstein metrics on compact Kähler manifold*, Invent. Math. 81 (1985), 359-372.

[Co] K. Corlette, *Flat G-bundles with canonical metrics*, J. Diff. Geom. 28 (1988), 361-382.

[D] S.K. Donaldson, *Connections; cohomology and the intersection forms of 4-manifolds*, J. Diff. Geom. 24 (1986), 275-342.

[DFr] S. Donaldson and R. Friedman, *Connected sums of self-dual manifolds and deformations of singular spaces*, Nonlinearity 2 (1989), 197-239.

[Do] R.-T. Dong, Ph.D. Thesis, UC San Diego, 1990.

[DF] H. Donnelly and C. Fefferman, *Nodal sets for eigenfunctions of the Laplacian on surfaces*, Jour. of AMS, 3 (1990), 333-353.

[FKM] D. Ferus; H. Karcher and H. Münzner, *Cliffordalgebren und neue isopara-metrische Hyperflächen*, Math.Z. 177 (1981), 479-502.

[F] A. Freire, *On the Martin boundary of Riemannian products* J. Diff. Geom. 33, 1991, 215-232.

[Fu] H. Fujimoto, *Modified defect relations for the Gauss map of minimal surfaces*, J. Diff. Geom. 29 (1989), 245-262.

[GY] L.Z. Gao and S.T. Yau, *The existence of negatively Ricci curved metrics of three manifolds,* Invent. Math. 85 (1986) 637-652.

[G] Gerhard, unpublished.

[HS] R. Hardt and L. Simon, *Boundary regularity and embedded solutions for the oriented Plateau problem*, Ann. Math. 110 (1979), 439-486.

[H] N. Hitchin, *On the curvature of rational surfaces, Differential Geometry* (Proc. Sympos. Pure Math., Vol. 27, Part 2, 1973), 65-80.

[HL] W.Y. Hsiang and H.B. Lawson, Jr., *Minimal submanifolds of low cohomogeneity*, J. Diff. Geom., 5 (1971), 1-38.

[Hu] Huisken, unpublished.

[Ja] P.S. Jang, J. Math. Phys. 141 (1976).

[J] L. Ji, *Spectrum degeneration for hyperbolic Riemann surfaces*, to appear in J. Diff. Geom.

[Jo] Jorgenson, to appear.

[JY] J. Jost and S.T. Yau, *The strong rigidity of locally symmetric complex manifolds of rank one finite volume*, Math. Ann. 275 (1986), 291-304.

[K] Y. Kawamata, *On deformations of compactificable complex manifolds*, Math. Ann. 235 (1978), 247-265.

[LT] P. Li and L.F. Tam, *Positive harmonic functions on complete manifolds with non-negative curvature outside a compact set*, Annals Math. 125 (1987), 171-207.

[Li] C.S. Lin, *The local isometric embedding in R^3 of 2-dimensional Riemannian manifolds with nonnegative curvature*, J. Diff. Geom. 21 (1985), 213-230.

[Me] W. Meeks, *III. Recent progress on the geometry of minimal surfaces in $\mathbb{R}^3$ and on the use of computer graphics as a research tool*, Proc. of I.C.M., 1986, 551-560.

[MY] W. Meeks and S.T. Yau, *The topology of three manifolds and the embedding problems in minimal surface theory*, Annals of Math. 112 (1980), 444-484.

[MM] M.J. Micallef and J.D. Moore, *Minimal two-sphere and the topology of manifolds with positive curvature on totally isotropic two-planes*, Annals of Math. 127 (1988), 199-227.

[Mo] S. Mori, *Threefolds whose canonical bundles are not numerically effective*, Ann. Math. 116 (1982), 133-176.

[M1] H.F. Münzner, *Isoparametrische hyuperflachen in spharen*, Math. Ann. 251 ;(1980), 57-71.

[M2] H.F. Münzner, *Uber die zerlegung der sphare in Bal bundel II*, Math. Ann. 256 (1981), 215-232.

[O] R. Osserman, *A survey of minimal surface*, Van Nostrand Reinhold, New York, 1969.

[OT] H. Ozeki and M. Takeuchi, *On some types of isoparametrische Hyperflachen*, Math. Z. 177 (1981), 479-502.

[P] M. Perry, Gravitational instantons. pp. 603-630. Seminar on Differential Geometry (ed. by S.T. Yau), Princeton Univ. Press, 1982.

[PR] J. Pitts and J.H. Rubinstein. *Applications of minimax to minimal surfaces and the topology of 3-manifolds*, Mini conference on geometry and PDE, 2 (Canberra, 1986), 137-170. Proc. Centre Math. Anal. Austral. Nat. Univer. 12, Austral. Nat. Univ., Canberra, 1987.

[Sc] R. Schoen, *Recent progress on non-linear problems in geometry*, To appear in Survey in Differential Geometry (published by J. Diff. Geom., 1991)

[SY] R. Schoen and S.T. Yau, *The structure of manifolds with positive scalar curvature*, Directions in PDE, Academic Press, 1987, 235-242.

[Si] C. Simpson, Ph.D. Thesis, Harvard, 1987.

[S1] B. Solomon, *The harmonic analysis of cubic isoparametric minimal hypersurfaces I; dimension 3 and 6*, Amer. J. Math. 112 (1990), 157-203.

[S2] B. Solomon, *On the Gauss map of an area-minimizing hypersurface*, J. Diff. Geom. 19 (1984), 221-232.

[St] S. Stoltz, to appear in the proceeding of AMS conference in Los Angeles.

[T1] G. Tian, *On Kähler-Einstein metrics on certain Kähler manifolds with* $C_1 > 0$, Invent. Math. 89 (1987), 225-240.

[T2] G. Tian, *On Calabi's conjecture for complex surfaces with positive first Chern class* Invent. Math. 101 (1990) 101-172.

[T3] G. Tian, *Smoothness of universal deformation space of compact Calabi-Yau manifolds and its Peterson-Weil metric*, pp. 629-646, Mathematical Aspects of String Theory (ed., S.T. Yau), World Scientific, Singapore, 1987.

[TY] G. Tian and S.T. Yau, *Three dimensional algebraic manifolds with* $a = 0$ *and* $X = -6$, pp. 543-559. Mathematical Aspects of String Theory (ed., S.T. Yau), World Scientific, Singapore, 1987.

[To] D. Toledo, to appear in the proceeding of AMS conference in Los Angeles.

[U] K. Uhlenbeck, *Removable singularities in Yang-Mills fields*, Comm. Math. Phys. 83 (1982), 11-30.

[UY] K. Uhlenbeck and S.T. Yau, *On the existence of Hermitian-Yang-Mills connections in stable vector bundles*, Comm. Pure Appl. Math. 39 (1986), 257-293.

[We] Wente, H., *Counterexample to a conjecture of H. Hopf*, Pacific J. Math. 121 (1986), no. 1, 193-243.

[W] R. Wentworth, *The asymptotics of the Arakalor-Green's function and Faltings' delta invariant*, Comm. Math. Phys. 137 (1991) 427-459.

[Wi] E. Witten, *New issues in manifolds of SU(3) holonomyu*, Princeton Preprint.

[Wil] P. Wilson, *Calabi-Yau manifolds with large Picard number*, Invent. Math. 98 (1989), 139-155.

[Wo] S. Wolpert, *Spectral limits for hyperbolic surfaces I and II*, Preprint, 1990 and 1991.

[X] F. Xavier, *The Gauss map of a complete non-flat minimal surface cannot omit 7 points of the sphere*, Annals Math. 113 (1981), 211-214.

[Y] S.T. Yau, *Uniformization of geometric structures*, Proc. Symp. Pure math., H. Weyl memorial volume.

[Z] F. Zheng, Ph.D. Thesis, Harvard University, 1989.

My Associations With S.S. Chern

Chuan-Chih Hsiung
Lehigh University

A Chinese translation of the well-known book in honor of S.S. Chern's eightieth birthday will soon be published. I was pleased to be asked by S.T. Yau to contribute this personal account of my interactions with Chern.

There is no doubt that Chern is one of the great geometers of the twentieth century. He has made unique and significant contributions to the research and development in the field of differential geometry. His research covers very broad areas, not only global, but also local; his portfolio of publications contains many important papers. In my opinion, his greatest contribution is still the simple proof of the Gauss-Bonnet formula published in 1944. He obtained the proof by applying E. Cartan's exterior differentiation (which, at that time, was not well understood by most people) and the transgression. This proof was such an outstanding new work in that period that H. Hopf stated that differential geometry had entered into a new era.

In fact, in 1946, Chern, using the same method, obtained the surprising and enduring Chern class. As an aside, he told me that before the discovery, he also wanted to use those curvature two-forms to express the Pontrjagin class; however, he could not succeed and obtained some unfamiliar and unexpected results. Since at that time he did not know that the original goal was impossible, he was disappointed and for a while did not even want to write up these unusual results. However, he had spent so much time on this research, that after much thought, he decided to publish them. So, he sent the paper to the Annals of Mathematics where it was immediately published. At that time, no one realized that these new unexpected results would become the Chern class. This incident supports the belief that new creations are not easy to achieve.

History

Chern went to Nankai University in 1926 when he was only fifteen years old. In 1920, Lifu Jiang had established the Department of Mathematics there and in a few years, had built one of the most important mathematics departments in China. This department at Nankai University has nurtured many future talents. It was in this environment that Chern received four years of solid and fundamental training which would contribute to his later development.

As an aside, in 1932 when I was a student at Zhejiang University, I learned from Professor Bao-Zung Chien (a famous Chinese mathematics historian who was from the same district as Chern and was good friends with his father) that Chern almost did not enter Nankai University because his father thought his son was too young to be so far away from home. Moreover, Nankai would be more expensive than a national university. So, Chern's father had decided not to let Chern go to Nankai and asked Chien to recommend his son to a national university. However, Chien had taught at Nankai for two years and knew its high standards; he finally convinced Chern's father to send Chern there. If the outcome had been different and Chern had attended a national university, who knows if Chern would have still developed into such an eminent differential geometer.

After graduation in 1930, Chern went to Qing Hua University to do research with Guangyuan Sun in projective differetial geometry; he subsequently wrote several papers. At that time, research in China was not progressive and Chern was one of the very few people who had not gone abroad and could publish research papers; of course, the mathematical community in China had already started to pay much attention to him.

My Associations With S.S. Chern

In 1932, I was studying at Zhejiang University. Since at that time the communication in China was very poor, especially since I was in the South and Chern was in the North, I was not able to meet Chern and could only admire his work. Our paths were to cross at many points before I actually met him.

Chern went to study in Hamburg, Germany in 1934 and returned to China in 1937. I had graduated and was doing research at Zhejiang University. Due to the Sino-Japanese war, he went to Southeast Associated University in Kunming whereas I followed Zhejiang University as it moved from Ishan to Guangxi to Tsunyi and Meitan in Guizhou. I still remember that Chern sent his paper on some invariants of contact of two curves in an n-dimensional projective space to my teacher Buchin Su; as a result, I had a chance to read it. Because of the war, the postal communication between China and other countries had stopped; so most mathematicians in China did not know that Chern's paper on the Gauss-Bonnet formula had already been published. I was aware of this paper because my teacher Su told me that Chern's former teacher Lifu Jiang had informed him of the news. Although I knew it was an important paper, my mathematics was still not mature enough then to understand its magnitude.

In 1946, I went to Michigan state to work on my PhD. That same year, Chern left Princeton to return to China to organize the Institute of Mathematics in Academia Sinica, the Chinese National Academy. Since I had just arrived in the US and was involved in my work at Michigan State, I missed another opportunity to meet Chern. Later, I met many people in the mathematical community in the US, especially P.A. Smith of Columbia University and A.A. Albert of the University of Chicago who expressed to me that they were interested in inviting Chern to teach, but he wanted to return to China. Albert further mentioned that Chern had an open invitation to the University of Chicago.

Due to changes in the political conditions in China, Chern (this time together with his family) came back to the US in 1949. He initially stayed at the Institute of Advanced Study in Princeton for the spring semester and then started to teach at the University of Chicago from the beginning of the following summer. Since I was an instructor at the University of Wisconsin in Madison in 1950, I went to the meeting of the American Mathematical Society at the Illinois Institute of Technology. I was very happy to finally get a chance to meet Chern. It was of particular significance to me that after all our mutual contacts in China, our first meeting was outside China. From that time on, whenever my family and I were passing through Chicago, we went to visit Chern where we were welcomed and entertained by Mrs. Chern. Academically, Chern and I often went together to attend various scientific conferences (including, of course, geometry) sponsored by organizations such as: American National Science Foundation, American Mathematical Society, University of Warwick in England, National Mathematics Research Institute of West Germany at Oberwolfach, Canadian Mathematical Society and other institutions worldwide.

In 1950, I went to teach at Northwestern University and then went to Harvard University on a research fellowship. When G. Birkhoff was chairman of the Department of Mathematics, he came to my office one day and said to me that since I was there, he wanted to invite Chern to come to Harvard to give a one-month seminar on differential geometry. He asked my opinion and I thought it was an excellent idea and would increase the interest in differential geometry at Harvard. The next day, Birkhoff told me that the Department had approved the invitation and had arranged living accomodations for Chern and his family. He wanted me to take care of their transportation and since I had an old car, I could foresee no problems. During that month, our two families often spent time together. There were many Chinese in the Boston area; Chern knew a number of them, so we had a busy and lively social schedule. At that time, Harvard did not have any differential geometers. However, Singer and Ambrose were working in differential geometry at MIT and they took the opportunity to go to Chern often for instruction. This helped Chern feel less isolated in Cambridge.

In 1959, I obtained some uniqueness theorems on closed convex hypersurfaces in a Euclidean space. I sent the paper to Chern and asked him to recommend it to a journal for publication. At that time, he was also working in the same direction as my paper and considered my findings significant. As a consequence, Chern and J. Hano extended my results and the three of us published a joint paper. In 1960, Chern left Chicago for Berkeley. I expressed an interest in seeing him in his new environment and working with him for a short time. Chern was able to obtain a grant for me from the National Science Foundation, so in 1962, I worked with him at Berkeley for half a year. After I arrived, he wanted me to prove his conjectural isometric condition for two compact hypersurfaces in a Euclidean space. In three or four months, I was able not only to prove the condition, but also to extend the two hypersurfaces to any two submanifolds of the same dimension. This extension was unexpected and not anticipated beforehand.

I had always wanted to found a journal in differential geometry to facilitate the development of research in this field. This dream was continuously postponed because my research and teaching kept me so busy. Finally in 1966, I thought that I could no longer delay and began to work on establishing the journal for a 1967 initiation. The first person with whom I discussed the plans was Chern and asked for his opinion. He also considered the publishing of such a journal a necessity and agreed with me as how to go forward. I would be the principal manager and with his support and help from the other editors whom I enlisted, the first volume of the Journal of Differential Geometry was published in 1967, as originally planned.

In 1971, Chern was 60 years old and in 1972, another honorary editorial advisor, D.C. Spencer would also become 60 years old.

At that time, our journal was published four times a year. To celebrate both editors' birthdays together, I selected four consecutive issues, two issues in the last half of 1971 and two issues in the first half of 1972.

Chern appreciates Chinese classic poetry and sometimes writes poetry for his own pleasure; I also like to read and write Chinese poetry. To celebrate his sixtieth birthday, I composed a classical four line, seven character poem which was especially set on paper by an expert calligrapher. I sent the poem to Chern who told me that he liked my poem and hung the calligraphy in his study in his home, so that he could often look at it. I have literally translated the poem as follows:

```
Long admiring your unique talents,
but delayed in making your acquaintance
Very happy to go out often with you
after we met outside of China
Now you are still a tireless teacher
and researcher as in old times
A great master
with the everlasting celebrated Chern class
```

In 1990, I was invited to attend the conference honoring Chern's seventy-ninth birthday which was held in Los Angeles. Time passes so quickly and it is already six years ago. But Chern is still "a tireless teacher and researcher as in old times". I admire him very much. I will stop my reflections here and wish him longevity.

BECOMING AND BEING A CHERN STUDENT

Thomas Banchoff
Brown University

Professor Chern arrived at Berkeley in 1960, the same time that I began there as a graduate student, and like most people in the department, I attended the first lecture of his graduate course in differential geometry, a subject I knew almost nothing about. I only lasted one lecture, not because I didn't understand what he was talking about but because I thought I did. In my senior year at Notre Dame, I had just spent a full semester in an independent study course on convexity and he covered everything that I had learned in about a quarter of an hour. The message was clear to me – this was not a course where I should try to learn about differential geometry for the first time. I enrolled instead in the undergraduate course on curves and surfaces using Struik's classic text, and I fell in love with the subject. I looked forward to taking Chern's graduate course the following year.

I was disappointed that it was not Chern who taught the graduate course the next year, but rather a visiting professor. It was a very good course, and I found that I was asking somewhat different questions from my fellow students, and occasionally coming up with a different solution to a difficult problem, a good sign that I was finding my way in this new subject. But I was still sorry I missed the chance to take that course from Chern.

My opportunity to meet Chern came in the summer of 1962. In my first two years, I had studied a good deal of geometry and topology, and after I got my qualifying exams out of the way, Ed Spanier invited me to stay on and begin some research under his contract. That summer the American Mathematical Society ran a large conference on Differential Geometry and Relativity at UC Santa Barbara. John Archibald Wheeler and Chern were scheduled to give the initial lectures, and I drove down to hear them. The physics went right by me, but the mathematics was wonderful. I was in the habit of taking very complete lecture notes and I wrote down just about every word Chern said. The lecture was a tour de force, starting with the sum of the angles of a triangle and going all the way to the Chern-Weil homomorphism in one dense hour. It was exhilarating.

After the lecture, several of the Physicists came to Chern and asked if they could see his notes. They were sure that they could master all he had said if they could look at some notes for another half hour or so. When he said he did not have any, someone mentioned that he had noticed a graduate student who seemed to be taking very good notes – maybe they could be duplicated? (I was reminded of the Gospel story where the Apostles were wondering how to feed the multitude and someone said that there was a young boy with few loaves and fishes).

I was summoned to show my notes to Chern. "Bring them around tomorrow morning", he said, "and I'll take a look at them". I certainly worked very hard that night! With the help of my classmate Robby Gardner and a couple of other graduate students, I filled out some of the missing steps in some of the faster parts, so the next morning I was able to present a clean and somewhat detailed version to Chern. He liked it. He added section headings and then crossed out the details I had filled in, explaining that it was important to leave some things for the reader to do. Those mimeographed notes became part of the official record for the conference. The AMS director Gordon Walker invited me to stay for the entire three weeks, with all expenses paid, and I started to think of myself as a geometer.

At the end of the conference, Chern asked me if I would take notes for his graduate topics course this fall. I said I would be very happy to do that, and he said, "Good. I will tell Spanier that you will be moving over to my research contract." So that was that.

During that course, Chern had students give lectures on current literature and he asked me to report on a paper of Louis Nirenberg "On a Class of Closed Surfaces". The main result was a rigidity theorem, showing that two isometric surfaces with minimal total absolute curvature were necessarily congruent. I duly presented the result, after a great deal of preparation and considerable help from other graduate students more versed in ordinary and partial differetial equations. The differential equations methods required some technical non-geometric assumptions, and I was determined to try to remove them. Since minimal total absolute curvature was a generalization of convexity, I decided to try to approach the problem using the synthetic and polyhedral methods of A.D. Alexandroff and Pogoreloff. I found a geometric way to describe the total minimal absolute curvature condition that worked for polyhedral as well as smooth surfaces, and I set out to prove a rigidity theorem for polyhedral surfaces satisfying this condition. Although this was not the kind of method that Chern had used, he encouraged me in this direction, and I began to report my progress to him each week.

After working several months and proving some small results with strong hypotheses, I surprised both Chern and myself by discovering a counterexample – I constructed two isometric but non-congruent polyhedral tori, both satisfying the minimal total absolute curvature condition! I felt that the whole project had fallen apart.

Chern suggested a change of topic. I began to try to generalize in higher dimensions some theorems Robert Osserman had proved for surfaces using techniques from one complex variable. I appeared to make progress at first, but six months later I had still no substantial results. Then one Saturday morning I got a call from Chern asking if I could come right up to his home. That had to be serious. He told me he had just come back from a conference where he had heard that Osserman and Robert Finn had already found a generalization, and when we looked at their methods and their results, it was clear that my approach was not going to work at all. It was another complete thesis disaster.

Fortunately there was little time for me to become too discouraged. Just weeks later, Chern called me to his office to meet a visiting mathematician. "You two think alike", he said and he introduced me to Nicolaas Kuiper. When I brought out my non-congruent isometric polyhedral tori, Kuiper showed interest. He asked, "Perhaps you would like to prove polyhedral analogues of some of the theorems on total absolute curvature in my recent papers?" His favorite result at that time stated that any smooth surface in six dimensional space with minimal total absolute curvature had to be contained in an affine hyperplane.

For the next two weeks, I worked hard trying to show that any polyhedral surface in six-space satisfying the minimal total absolute curvature condition would have to lie in a hyperplane. Then one day as I was folding the wash in the local laundromat, I discovered that that conjecture was not true. It was possible to find such an example in six-space, and in fact I could construct such polyhedral examples in arbitrarily high dimensions!

I made some folded out paper models with instructions how to assemble them in six-space, and the next day I showed them to Kuiper. He was very surprised. He said to me, "What you have here is a gold mine. I'll give you six months. If you haven't written a thesis by that time, I'll give the problem to one of my own students. It's too good a problem not to be solved by someone."

As it happened, I didn't need six months. Chern was very supportive of this new project, although the methods were very far from the geometry he was working on at the time. A few weeks later in our weekly meeting, I presented a small result with some very restrictive hypotheses and conjectured that it should be true in general. Chern made a comment that commited him – he said that my conjecture was quite ambitious and he would be surprised if I could prove it. When I was able to prove it, the thesis was well on its way.

Chern's guiding hand made another big difference at this crucial point. I was very excited by the topic and proposed grand theories of polyhedral critical point theory and generalizations of curvature in high dimensions. Chern advised me to write up my basic results and to put the general theory aside: "In a thesis, you don't have to say the last word on the topic. A thesis is supposed to prove that you are able to do research. A few years from now you will understand the general problem better."

He was right of course. With his encouragement, I became a Benjamin Pierce Instructor at Harvard, and then I spent a posdoctoral year with Kuiper in Amsterdam. I was happy to send him my paper on the general theory of critical points and curvature for embedded polyhedra. It was much better than I could have done at the time I was finishing my thesis.

Chern also played a central role in shaping the part of my career that deals with computer graphics. Shortly after I came to Brown University in 1967, I began working with a computer science colleague Charles Strauss to create computer generated animated films of surfaces in four-dimensional space. By the time of my sabbatical in 1974, we had produced several films, including the Flat Torus, the Hypercube, and Complex Function Graphs. I spent the first quarter of my sabbatical at Berkeley, and Chern came to see the films when I presented them in two colloquia. Two years later, he saw our further work on Evolute Surfaces of Ellipsoids and the Veronese surface. He said then that he thought it was time to present such films to the mathematical world at the International Congress of Mathematicians to be held in Helsinki in 1978. He suggested the possibility to Armand Borel, and I came to the Institute at Princeton for a "command performance". The audition was successful, and I gave an invited talk at the Congress, with a repeat presentation later in the week at the request of the organizers. It was a high point in my mathematical life, thanks to the faith Chern had in me.

Over the years, whenever I have come back to Berkeley, I have always enjoyed visiting with Chern and with Mrs. Chern. His great kindness and his encouragement have supported me at every stage of my career, and I will always be deeply grateful and proud that I am a student of Professor Shiing-Shen Chern.

Textbooks from International Press

Calculus: A Computer Algebra Approach,

by L. Ashel and D. Goldfeld

Uniquely designed for use with computer algebra systems and sophisticated calculators, this course also works well with a computer laboratory. The students are encouraged to use technology for manual computation while they rapidly progress through the concepts of differential and integral calculus, mathematical modeling and optimization, ordinary differential equations, differential calculus for vector valued and multi-variable functions. The students will progress to vector geometry and coordinate systems, two and three dimensional graphical display, multiple integration, vector fields and line integrals, and on to Fourier series and the Fourier expansion theorem.

Basic Partial Differential Equations,

by D. Bleecker and G. Csordas

Using partial differential equations as a tool to predict systems based on underlying physical principles, this text is accessible to students with calculus background without presupposing courses in linear algebra or ordinary differential equations. The text takes an inductive approach, motivating students to see the relevance and applications of the general principles which govern the subject, but rigorous proofs are still available.

The problems included range from routine to very challenging. The problems are constructed to require genuine understanding of principles as well as procedures, and the solutions manual is available only to instructors—not to the general public. Suggestions from the author indicate how the book can be used in one semester, two quarter, or two semester courses.

A First Course in Differential Geometry,

by C.-C. Hsiung

This text treats the traditional topics of curves and surfaces in 3-dimensional Euclidean space with attention equally on local and global properties. Extension to higher dimensional spaces and more general surfaces is available. In addition to analytical approaches, the text gives geometric interpretation to its subjects and encourages the development of geometric intuition. The book uses vector analysis and exterior differential calculus but does not require tensor calculus.

Part 1 reviews point-set topology, advanced calculus, and linear algebra. Part 2 establishes a general local theory for curves in three dimensional Euclidean space including proving the uniqueness theorem for curves in E^3. Part 3 develops a Local theory for surfaces in E^3, and Part 4 covers orientation of surfaces, surfaces of constant Gaussian curvature, and global theorems for surfaces.

Proceedings of the International Seminar on Singularities and Complex Geometry, *edited by Qi-Keng Lu, Stephen Yau and Anatoly Libgober*

Surveys in Differential Geometry, Volume I, Volume II, *edited by C.C. Hsiung and S.-T. Yau*, **Volume III,** *edited by Chuu Lian Terng and Karen Uhlenbeck.*

Topics in Symplectic 4-manifolds, volume 1, First International Press Lecture Series, *edited by Ronald Stern*

Tsing Hua Lectures on Geometry and Analysis, *edited by S.-T. Yau*

Current Developments in Mathematics

Proceedings of the annual conference sponsored by Harvard and MIT

1995 Proceedings: lectures by Henri Darmon, Fred Diamond, and Richard Taylor, Mikhail Lyubich, Ib Madsen, Curtis McMullen, and Gang Tian

1996 Proceedings: lectures by Richard Borcherds, Gerrit Heckman and Eric Opdam, Ehud Hrushovski,k and Yves Meyer

1997 Proceedings: lectures by A. Connes, L. C. Evans, P. Sarnak, and W. Soergel and open problems by A. Jaffe, D. Stroock, B. Mazur, C. Taubes, and D. Jerison

Mathematical Physics Books from International Press

www.intlpress.com
PO Box 38-2872, Cambridge, MA 02238-2872

Quantum Groups: from Coalgebras to Drinfeld Algebras *by Steven Shnider and Shlomo Sternberg*

75 years of Radon Transform *by Simon Gindikin and Peter Melchior*

Perspectives in Mathematics Physics *by Robert Penner and S.T. Yau*

Proceedings of the International Conference on Mathematical Physics XIth volume edited by Daniel Iagolnitzer, XIIth volume edited by David DeWitt

Mirror Symmetry volume 1 edited by S.T.Yau, volume 2 edited by Brian Greene

Physics Series

Physics of the Electron Solid *by S.-T. Chui*

Proceedings of the Second International Conference on Computational Physics, *by D.H. Fenb and T.-Y Zhang*

Chen Ning Yang, a Great Physicist of the Twentieth Century, *by S.-T. Yau*

Yukawa Couplings and the Origins of Mass, *edited by P. Ramond*

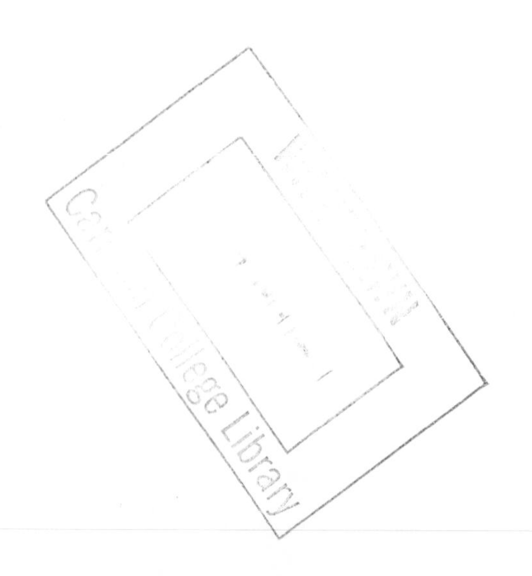